Springer Theses

Recognizing Outstanding Ph.D. Research

Aims and Scope

The series "Springer Theses" brings together a selection of the very best Ph.D. theses from around the world and across the physical sciences. Nominated and endorsed by two recognized specialists, each published volume has been selected for its scientific excellence and the high impact of its contents for the pertinent field of research. For greater accessibility to non-specialists, the published versions include an extended introduction, as well as a foreword by the student's supervisor explaining the special relevance of the work for the field. As a whole, the series will provide a valuable resource both for newcomers to the research fields described, and for other scientists seeking detailed background information on special questions. Finally, it provides an accredited documentation of the valuable contributions made by today's younger generation of scientists.

Theses are accepted into the series by invited nomination only and must fulfill all of the following criteria

- They must be written in good English.
- The topic should fall within the confines of Chemistry, Physics, Earth Sciences, Engineering and related interdisciplinary fields such as Materials, Nanoscience, Chemical Engineering, Complex Systems and Biophysics.
- The work reported in the thesis must represent a significant scientific advance.
- If the thesis includes previously published material, permission to reproduce this must be gained from the respective copyright holder.
- They must have been examined and passed during the 12 months prior to nomination.
- Each thesis should include a foreword by the supervisor outlining the significance of its content.
- The theses should have a clearly defined structure including an introduction accessible to scientists not expert in that particular field.

More information about this series at http://www.springer.com/series/8790

Abhijeet Patra

Quantifying Interactions of Biomolecules with Inorganic Surfaces

Doctoral thesis accepted by
the National University of Singapore, Singapore

Springer

Author
Dr. Abhijeet Patra
NUS Graduate School for Integrative Sciences and Engineering
NUS Nanoscience and Nanotechnology Institute
Singapore

Supervisor
Prof. T. Venkatesan
Department of Physics, Electrical Engineering, Material Sciences and Engineering, NUS Graduate School for Integrative Sciences and Engineering
NUS Nanoscience and Nanotechnology Institute
Singapore

ISSN 2190-5053 ISSN 2190-5061 (electronic)
Springer Theses
ISBN 978-3-319-80876-5 ISBN 978-3-319-30728-2 (eBook)
DOI 10.1007/978-3-319-30728-2

Printed on acid-free paper

This Springer imprint is published by Springer Nature
The registered company is Springer International Publishing AG
The registered company address is: Gewerbestrasse 11, 6330 Cham, Switzerland

Supervisor's Foreword

Biology, today, has become an increasingly interdisciplinary field of enquiry. The quest to understand the state of human health, identify the causes of diseases and subsequently design methodologies of diagnosis and treatment have brought chemists, physicists, engineers and traditional biologists to work together, striving to think beyond the box. As is often found to be the case in any field of science, integration of non-traditional skills and ideas with deep formal understanding leads to potential breakthroughs from completely unexpected directions.

Making quantitative and meaningful measurements have generally proven to be a difficult problem to tackle in biology and its allied domains. The sheer complexity of the biological systems leads to a worrying scenario where the possibility of artefacts being recorded as meaningful data is unacceptably high. Of course, bright minds have made impeccable progress over the years, seeking to establish proper protocols and pedagogy to make reproducible measurements and enable meaningful data.

In this piece of work, we have demonstrated how existing tools and techniques can be adapted to attack some of the relevant problems in biomedical sciences. Our strength has been the ability to make highly controlled experiments which has generated knowledge which is immediately applicable. For instance, we have used the interactions of the atomic force microscopy tip with a sample to predict its hydrophilicity. The result is a technique which can locally evaluate (across spatial resolutions in tens of square nanometers) the wettabilities and relative surface energies of any surface and have shown that the learning from such a small volume can be extended to macroscopic sample sizes.

Working with collaborator Prof. Markus Wenk, also at National University of Singapore (NUS), we took up the challenge of resolving a long-standing bottleneck in the area of sample preparation in lipidomics. The challenge was to design a solid phase extraction (SPE) resin which would be able to perform an unbiased cleanup of the sample matrix enabling the mass spectrometry (MS) systems to identify more lipid species. We ended up fabricating lipid extraction devices in a number of machine-compatible formats in addition to methods which enable accurate quantification of lipid adhesion to surfaces used as SPE resins. Furthermore, we discovered how some materials which are, in fact, in use today can contribute to ambiguous results due to their leaching under sample preparation conditions.

The last problem we have aimed to tackle is the study of the protein corona that forms around nanoparticles once they are exposed to the serum environment. In collaboration with Assistant Professors James Kah and Chester Lee Drum at NUS and Professor Bo Liedberg at Nanyang Technological University, we developed a technique based on multiplexed surface plasmon resonance which make it possible to observe, in real time (first report of its kind), the interactions between nanoparticles and serum proteins. The method demonstrated takes 3–4 hours to return kinetic parameters of serum interactions with nanoparticles, which in turn, make it possible to calculate approximate in vivo surface coverage. Competing technologies are either incapable of delivering the same knowledge, or are far less accurate or take significantly far more time to provide similar information. We expect the technique to be readily adopted by the nanomedicine community because of its promise.

This thesis demonstrates what is possible when a graduate student has the courage to think out of the box and manages to collaborate with researchers across numerous field boundaries to produce novel breakthroughs.

Singapore
October 2016

Prof. T. Venkatesan

Abstract

Making precise and meaningful measurements of interactions in biological sciences have traditionally proven to be challenging. The complex nature of experiments, innumerable interferences and the sheer manual efforts required have made it a formidable, but extremely interesting problem to solve. In this thesis, we have tried to develop techniques which enable measurements where previously, none existed and make measurements more accurate or convenient or both, where they were erroneous or cumbersome or both.

The ability to quantify the interactions between biomolecules and synthetic surfaces are crucial to our knowledge of several key biochemical scientific domains. We begin by taking a closer look at the simplest biomolecule, water. Interactions of water with surfaces have been studied mostly through the century-old method of water contact angle and the subsequent derivation of surface energy. However, in today's age of nanotechnology, this method has been dealt a severe limitation because of its macroscopic nature. Measuring local surface energies over areas no larger than hundreds of nanometres is very important today in the context of solid-liquid interfaces, nanofluidics, photovoltaics, biologics and several other allied domains. We demonstrate a new method based on force spectroscopy, to locally probe contact angle and surface energies, in the nanometre scale. The results we have obtained correlate very well to the macroscopically measured values.

Subsequently, we studied interactions of phospholipids with metal oxide surfaces with a close view on fabricating better solid phase extraction (SPE) resins to improve lipid fraction enrichment. Sample preparation in lipidomics remains, till date, a major bottleneck, in the way of discovering underrepresented lipid species within the lipidomes of organisms. The overwhelming presence of phospholipids suppress the ionization and consequently the discovery of all rare lipids. We evaluate the performance of metal oxides as SPE resins to remove phospholipids from lipid extracts. We also demonstrate a very sensitive method of quantifying phospholipid adhesion to metal oxide surfaces using the phenomenon of surface plasmon resonance. Thereafter, we use the knowledge gained to fabricate enrichment devices in various formats.

Finally, we study interactions of plasma proteins with the surface of modified and unmodified nanoparticles. Non-specific adsorption of serum proteins onto the high energy surface of a nanoparticle which hits the physiological environment has been the biggest problem in the way of understanding and designing nano drug delivery systems. This adsorbed layer of proteins, also called "protein corona" completely alters the identity of the nanoparticle and radically changes its pharmacodynamics and pharmacokinetics. Existing methods to study and monitor the protein corona have been found to be either rudimentary or cumbersome. We demonstrate a surface plasmon resonance based high-throughput screening protocol to study the protein corona formation on nanoparticles. This method enables the user to derive kinetic constants of interaction which in turn help model the *in vivo* assembly of proteins on the nanoparticle surface. Ultimately, we show that the technique can be used to screen through nanoparticle based drug delivery formulations on few key parameters such as resistance to biofouling and ability to find intended target, within 3–4 hours thereby saving significant amounts of time and money.

Acknowledgement

I do not believe in providence, fate or destiny. I believe that our present efforts have repercussions and step by step, they set out ripples which construct our future. Every one of our past actions shape our present through these repercussions, some severe, some gentle. However, when I think of how I came to know of Venky (Prof. T. Venkatesan, as he is formally known) and decided to join his group to pursue my doctoral research, I am stumped. I can't think of anything I have done in the past which could result in my walking into his office on 3 November 2011. That is probably what has been known as chance or luck. What has happened thereafter is, in retrospect, simply amazing, even borderline magical.

An airplane is very safe in a hangar. But that is not what an airplane is for. Under Venky's guidance, I learnt how to fly. Without Venky's zeal for pushing the boundaries for knowledge and constant lobbing of challenges at me, my graduate life would have been boring, perhaps even frustrating. Venky dislikes mediocrity and he has trained me to be extremely uncomfortable with it. Here was a highly accomplished world-beating scientist, 65 years of age, asking questions with a curiosity of a first year undergraduate student. In equal parts, Venky has amazed and taught, in the process lighting an indomitable fire within me. I feel very proud to say that he has imbibed in me the true spirit of scientific enquiry. A very large part of the success I get in my life hereafter will have its roots in this time which I have spent under Venky's supervision. I can't thank you enough, Venky. But, I sure will make you very proud.

My parents and my elder brother have always been scared of the choices I have made in life so far. In spite of their fear, they have always let me choose what I wanted to do and supported me in my decisions. I owe everything I am today, to your love and support *Ma, Baba* and *Dadabhai.*

There is another person who I need to specially thank. Her name is Ritika "Motu" Borkar. She kept me on track when the going got tough. She never let my self-belief slide always telling me that I was going to do great, repeating it until I started believing it again. She went to all lengths to ensure that I was happy, excited and comfortable in life. For being who you are and for caring as much as you do, I do not have words but ineffable feelings. Thank you, Motu.

I would like to thank Dr. Federico Tesio Torta for being my mentor and teaching me the ropes of lipidomics research. He has been extremely patient with me and immensely helpful. Working with him has been an extremely rewarding experience. Dr. Pradeep Narayanaswamy has been a friend and has helped me with his mass spectrometry expertise. Thanks to Aveline for helping me in the initial days of lipidomics research. I also have to thank Prof. Markus Wenk for his inputs on the development of a solid phase extraction resin. Dr. Wang Yi has helped me a lot with surface plasmon resonance and really accelerated my learning. I need to thank Prof. Bo Liedberg for his crucial insights and sharing his vast experience in the field with me. I am grateful to Dr. Tao Ding for his role in getting protein-nanoparticle interactions off the ground. I am also thankful to Prof. James Kah, Prof. Chester Lee Drum and my friend, Gokce. I also have to thank Kingshuk for his pointers in the initial days of my research. I am grateful to Prof. CT Lim and Prof. Ding Jun for their inputs on my research in their role as members of the thesis advisory committee.

Life in graduate school would not have been memorable without friends. I have made a lot of lasting friendships during my four years at NUS. These friends have not only made the last leg of student life exactly as crazy as it should be, but also been tremendous pillars of support and inspiration. I can't begin to thank all of them enough. First, I have to thank Michal Dykas for introducing me to Venky. Then I have to thank him again for being one of the best friends I have made in life, introducing me to Polish culture and the extremely complicated Polish language. "Papa" Lokesh Dhakar has been my other friend in need. He has played various roles—friend, financial saviour, startup advisor. From discussing general frustrations to intoxicated ponderings of the meaning of human life, Papa has been omnipresent. I have to thank Prerna for just being her entertaining self, Alok for all the musical inspiration and Arunabh for all the movies, food and entrepreneurial companionship.

I have to thank Dr. Siddhartha for his words of wisdom in times of need and his stimulating discussions on economics, culture and politics. I have learnt a lot about general conduct from Dr. Surajit, apart from his in-depth knowledge on spectroscopic techniques. Thanks to Dr. Malli for taking out lots of time to teach me the operation of the extremely complicated Rutherford Backscattering setup at Centre for Ion Beam Applications.

I would like to thank my friends at NUSNNI Nanocore—Dr. Abhimanyu, Dr. Meenakshi, Changjian, Tarapada and Soumya. There are two friends who I need to acknowledge separately for their contribution to this work. They joined our group in the last year of my candidature but have become close friends in a very short time. One of them is Sreetosh "Gurudeb" Goswami. He is a genius and a true epitome of devotion to one's pursuit. I have learnt a lot from him, least of which is to have single-minded dedication to unravelling the mysteries of experimental data. Almost every discussion with him is enlightening because of the sheer knowledge he can impart. The other is Saurav Prakash. Working with Saurav has been a complete joy. His understanding of material sciences is fantastic and the clarity with which he can explain concepts is admirable. We have banged our heads against the wall over seemingly nonsensical data together and then had eureka moments nonetheless.

This one is for all the crazy times, the ecstasy and the despair. Here's to the mysteries of science.

Contents

Abbreviations

AFM	Atomic Force Microscopy
ApoA1	ApolipoproteinA1
AuNP	Gold Nanoparticle
Cer	Ceramide
DFT	Density Functional Theory
DMPC	1,2-dimyristoyl-sn-glycero-3-phosphocholine
EDAC	1-ethyl-3-(3-dimethylaminopropyl) carbodiimide
ESI	Electrospray Ionization
FS	Force Spectroscopy
GluCer	D glucosyl-ß-1,1'-N-octanoyl-D-erythro-sphingosine
HPLC	High Performance Liquid Chromatography
HSA	Human Serum Albumin
IgG	Immunoglobulin G
K_A	Association Constant
K_D	Dissociation constant
k_{off}	Off-Rate
k_{on}	On-Rate
LAO	$LaAlO_3$
LC	Liquid Chromatography
MALDI	Matrix Assisted Laser Desorption Ionization
MS	Mass Spectrometry
NHS	N-hydroxysulfosuccinimide
OWRK	Owens Wendt Rabel and Kaelbel
PEG	Polyethylene Glycol
PI	Phosphatidylinositol diC16
PIP	Phosphatidylinositol 3-phosphate diC16
PIP_2	Phosphatidylinositol 3,4-bisphosphate diC16
PIP_3	Phosphatidylinositol 3,4,5-trisphosphate diC16
PLD	Pulsed Laser Deposition
RBS	Rutherford Backscattering Spectroscopy

RMS	Root Mean Square
RP-HPLC	Reverse Phase High Performance Liquid Chromatography
RU	Response Units
SEM	Scanning Electron Microscope
SM	N-lauroyl-D-erythro-sphingosylphosphorylcholine
SPE	Solid Phase Extraction
SPR	Surface Plasmon Resonance
STO	$SrTiO_3$
TEA	Triethylamine
TFA	Trifluoroacetic Acid
TLC	Thin Layer Chromatography
WCA	Water Contact Angle

Chapter 1
Introduction

> "When you can measure what you are speaking about, and express it in numbers, you know something about it".
> —Lord Kelvin, Electrical Units of Measurement, Vol 1, 1883-05-03.

Abstract This chapter is meant to serve as a brief description of the aims and contents of the research work which produced this thesis. We will introduce some of the problems that we have sought to tackle and how this work is going to help researchers in a variety of fields.

Long before the 19th century, the importance of quantification to progress in science was obvious even to ancient Greek philosophers. A contemporary of Socrates, Philolaus of Croton (in today's southern Italy) remarked "actually, everything that can be known has a number; for it is impossible to grasp anything with the mind or to recognize it without this." Time after time, leading scientific minds across eras have expressed their views on the role measurement plays in pushing the boundaries of human knowledge. To quote Lord Kelvin once again, "if you cannot measure it, you cannot improve it."

We believe, the science and technology that facilitates and enables measurement of parameters are the cornerstones of scientific advancements. However, when proper quantification has been hard to accomplish, science has taken a detour through qualitative, empirical and quasi-quantitative approaches. More often than not, the field of biology and allied fields with a biological context have had to deal with this problem. Unlike domains in engineering and solid-state physics, measurements in biology have been traditionally very difficult. The sheer complexity of biological systems have made data collection and consequent analysis a challenge. The difficulty notwithstanding, forward-thinking minds have made shocking advancements over the course of last few decades in making the field of biology more navigable with numbers to put on observations.

The motivation behind this thesis work is, as the title suggests, enable the quantification of interaction of biomolecules with inorganic surfaces. Biology, today, has spawned a whole array of fields of scientific research which draw on a

A. Patra, *Quantifying Interactions of Biomolecules with Inorganic Surfaces*,
Springer Theses, DOI 10.1007/978-3-319-30728-2_1

host of other major sciences to ultimately advance and benefit the human race. They are all highly specialized domains linked by one common thread, understanding the state of human health better and seeking to restore it in times of disease through specifically engineered interventions. In all of these fields, as in general science, knowing one's numbers is of paramount importance. Through this work, we demonstrate techniques which enable accurate measurements of parameters of interest to scientists involved in the study of lipids and proteins. In some cases, we have been able to measure quantities never measured before. In others, we have made measurements much more accurate and meaningful than reports in prior literature.

This thesis has been divided into 6 major chapters. The first chapter is an introduction to the thesis which describes the motivation behind taking up this work. We also discuss briefly the contents of the next chapters.

In Chap. 2, we will discuss the methods used to accomplish this study. The description and discussion is brief where the original established method has been used without any significant modification. A detailed and more in depth discussion is provided where the prevalent and commonly used method has been significantly tweaked to serve the purpose needed. We would like to draw special attention to the fabrication of the surface plasmon resonance chips as there are a number of issues which were taken care of and addressed in due course of this endeavor (please refer Sect. 2.5.4). We would also like to point out the modification made to the typical matrix assisted laser desorption ionization method as is prevalent in the field of proteomics and lipidomics with a detailed description of the process in Sect. 4.5.1.

For the sake of organization, the experimental work has been split into three major chapters. The class of biomolecules studied will increase in their weight from Chaps. 3 to 4 through 5. In seeking to quantify the interactions of biomolecules with inorganic surfaces, we begin with, in Chap. 3, the simplest and most common biomolecule, indispensable to the maintenance of life as we understand it, water. Thereafter, in Chap. 4, we will move to a heavier class of biomolecules, namely lipids which weigh in hundreds of Daltons. Chapter 5 will deal with interactions of proteins, which weigh in tens of thousands of Daltons. Since the functions and implications of each study are quite diverse and also not completely convergent, we decided against an overall literature review. Instead we have provided a detailed literature review of each section of study in the respective chapter. However, for the purposes of introducing the motivation behind taking up this work, we will describe the problems we have worked on and their importance to science in this chapter. We hope this makes it easier to explain the gaps in knowledge which we are trying to address through this piece of work.

The interactions of water with surfaces has been studied over hundreds of years and lots of advancements have been made in the field of surface and interfacial science. Wetting angle [1] as measured through the sessile drop method has been around for about a century. A slightly more complex measurement which enables inspection of hydrophilicity of a surface is inverse gas chromatography [2]. During the 1950s, scientists figured out how to use wetting angles to evaluate surface energies. These methods have been the only options for researchers looking to understand the interaction of water with the surfaces of interest. Of late, there has also been an explosion of interest in surfaces which would be intrinsically hydrophobic [3] for their applications in heat exchangers and high rise windows. The need to understand interaction of water with surfaces goes much further into fields of nanofluidics, photo-catalysis, solar energy conversion, self-assembled monolayers and many more. This brings us to the problem which existing methods mentioned above cannot solve. Almost all functional surfaces of today are nano-patterned (domains of different materials arranged on a surface). Contact angle measurement is a macroscopic measurement with droplet size usually in the range of μl. If we employ it in evaluating local wettability and consequently surface energies, the data obtained has inherent errors from chemical heterogeneity built into it. Engineers have tried to solve the issue by reducing the size of the droplet to nano and picolitres. However, making droplets that small introduces new problems—non-uniformity of sizes across successive droplets, quick evaporation from the surface, expensive equipment to dispense and manage such small droplets. Concurrently, groups around the world have tried to address the issue of wettability in the nanoscale by using simulation tools [4, 5]. However, upon extensive literature review (provided in detail in Sects. 3.2.1 and 3.2.2) we found that understanding of nanoscale phenomenon as gained through simulations hasn't been used to further understanding of macroscale interactions of water with surfaces, which is represented in contact angle. In Chap. 3, we demonstrate the ability to measure local surface energies with no more than an atomic force microscopy (AFM) setup. We analyse the traditional force-distance curves and find a direct relation between the force required to break free from the native water layer on any solid surface and its hydrophilicity. In essence, we show a new technique of quantifying interaction of water with unreactive inorganic surfaces. The technique we employ is simple and can be carried out within the most basic AFM setups. In order to really relate experimental data that we acquire to the accepted principles of the domain, we have also shown that there is a direct correlation between the propensity of a given surface to get hydroxylated (simulated through density functional theory calculations) with its tendency to be more hydrophilic (actual data collected through force-distance curves).

Chapter 4 deals with the interaction of phospholipids and transition metal oxides. Sample preparation has been a troublesome bottleneck in the field of lipidomics [6]. The mass representation of lipids species in the human lipidome is extremely skewed [7]. Mass spectrometry is the only analytical tool that researchers have to quantify these lipid species. Under this constraint, the overbearing abundance of phospholipids in the lipidome suppresses the ionization and subsequently discovery of underrepresented non-phospholipid species. An ideal solution would be the complete separation of phospholipids from non-phospholipids in the post-collection, sample preparation step. The goal is to analyze samples in a manner where one can gather signals from ionization, ideally from all species present. This task however is herculean. Due to the chemical diversity of the lipidome, there can't be one perfect strategy of preparing samples. There needs to be synergy between subsequent steps of sample preparation so that the final analysis discovers all species, without any bias being induced by any of the sample preparation steps. The latter part has proven especially difficult. More often than not, sample preparation strategies have not been completely understood and this has introduced the terrifying possibility of wrong conclusions being drawn from a compromised experiment [8–10]. Traditionally, sample preparation in lipidomics has revolved around liquid-liquid extraction, wherein the differential solubilities of a chemical species in an aqueous and organic phase is used to separate the species from the rest. However, applying this to phospholipid extraction from a lipidome is futile because almost all phospholipids prefer to stay in the organic apolar phase rather than in water. Emulsions are a real problem with liquid-liquid extraction as well. It is clear that the way forward through this bottleneck is definitely not through liquid-liquid extraction. Thereby, our aim comes to be defined as designing what is known in the field of chromatography as solid phase extraction resins. Solid phase extraction resins are materials, obviously in solid phase (powders or surfaces), which have selective affinity towards a class of chemical species (in this case phospholipids). Upon passing a complex lipidome through these resins, we should be able to selectively capture phospholipids on the resin, thereby leaving the flowthrough rich in non-phospholipid species. We wanted to develop accurate comparative studies between transition oxides with respect to their promise in being successfully employed as resins to perform selective phospholipid capture. There are reports in literature of various oxides [11, 12] being tried but measurements to compare between them hasn't been controlled to be able to draw the correct conclusions. When dealing with columns packed with powders of the intended resin, there are a few key parameters that one needs to keep in mind for gauging the performance of the column. Unless the particles of two resins being compared have the same size, the actual surface area in contact and the time spent by the separating species in contact will be different. Under, these circumstances, any choice made on the basis of such uncontrolled experiments, could prove to be dubious. Additionally, the resin material may corrode under the separation conditions being used, which might cause downstream effects in the workflow, especially when mass spectrometry systems are so sensitive to salts. We demonstrate a technique to precisely measure the amount of phospholipids captured per unit area of a thin film of transition metal oxide using surface plasmon resonance. The importance of being able to measure adhesion to atomically flat thin films is that in such a situation it is only chemistry of the surface that is influencing the binding. Such precise measurement would enable us to choose the best solid phase extraction resins for specific applications and optimize cost/performance.

In Chap. 5, we seek to quantify interactions between plasma proteins and gold nanoparticles. Lot of scientific effort has been directed towards understanding and quantifying the "protein corona" [13–15] which forms around nanoparticles when they enter the physiological environment. The non-specific adsorption of serum proteins onto nanoparticles has been blamed for a lot of engineered nano drug delivery systems failing to accomplish their task [16]. The accepted wisdom in the field is that the "corona" causes complete fouling of the surface thereby effectively covering all surface functionalization moieties which are intended to give it the ability to find and attach to specific cells, before releasing its cargo. The corona covered nanoparticle hence goes around the bloodstream or gets internalized by cells it is not intended for. Desirable clinical outcomes are completely diminished if not eliminated completely [17–20]. It is now known that silica nanoparticles [21], gold nanoparticles [22–25], carbon nanotubes [26, 27] all have different compositions of the corona that form on them. In a situation where engineered nanomaterials are being intensely investigated for their potential applications in healthcare, the study of protein corona formation is a very important yet very difficult problem. Upon review of literature (more details can be found in Sects. 5.2.2 and 5.2.3), it is revealed that there are two broad kinds of techniques being employed today. One group contains methods such as dynamic light scattering, differential centrifugal sedimentation, size exclusion chromatography, etc. which provide only the apparent change in radius when a nanoparticle binds to a protein creating a complex. In all these techniques, the data acquired varies greatly across runs and the information provided is just the apparent change in radius upon a binding event between the protein and the nanoparticle. If the unbound components are about the same size as the bound complexes, these techniques will fail to provide meaningful information. On the other hand, lies the most comprehensive method which provides all information regarding the identification of bound proteins to nanoparticles. However, the throughput is low and a large number of steps are involved which take about 8–10 days [28] to complete the measurement. The large number of manual steps also introduce the possibility of user bias. In addition, importantly, no method being used currently provides real time kinetics of the interaction between serum proteins and nanomaterials. We have addressed all these issues and designed a method which makes use of multiplexed surface plasmon resonance technique which provides kinetic information about protein adsorption onto nanoparticles in a matter of 3 h. Since our technique is modular, it can be used for any nanoparticle formulation of any shape, size within 100 nm. It can efficiently evaluate the efficacy of anti-fouling measures employed by formulation scientists to prevent corona formation. We also demonstrate how we can use this approach to screen nanoparticle based drug delivery systems in an ex vivo setup without having to do cell studies every time.

Once again, we wish to reiterate that a detailed survey of the related research landscape can be found at the beginning of each chapter which contains the actual experiments (Chaps. 3–5).

Chapter 6 contains concluding remarks and an overview of the future work which can follow the results which we have been able to showcase in this thesis.

References

1. Mittal, K.L.: Advances in Contact Angle, Wettability and Adhesion. Wiley (2013)
2. Das, S.C., et al.: Determination of the polar and total surface energy distributions of particulates by inverse gas chromatography. Langmuir **27**(2), 521–523 (2011)
3. Azimi, G., et al.: Hydrophobicity of rare-earth oxide ceramics. Nat. Mater. **12**(4), 315–320 (2013)
4. Giovambattista, N., Debenedetti, P.G., Rossky, P.J.: Effect of surface polarity on water contact angle and interfacial hydration structure. J. Phys. Chem. B **111**(32), 9581–9587 (2007)

5. Giovambattista, N., Debenedetti, P.G., Rossky, P.J.: Enhanced surface hydrophobicity by coupling of surface polarity and topography. Proc. Natl. Acad. Sci. **106**(36), 15181–15185 (2009)
6. Hu, C., et al.: Analytical strategies in lipidomics and applications in disease biomarker discovery. J. Chromatogr. B **877**(26), 2836–2846 (2009)
7. Quehenberger, O., et al.: Lipidomics reveals a remarkable diversity of lipids in human plasma. J. Lipid Res. **51**(11), 3299–3305 (2010)
8. Bruce, S.J., et al.: Investigation of human blood plasma sample preparation for performing metabolomics using ultrahigh performance liquid chromatography/mass spectrometry. Anal. Chem. **81**(9), 3285–3296 (2009)
9. Moco, S., et al.: Metabolomics technologies and metabolite identification. TrAC Trends Anal. Chem. **26**(9), 855–866 (2007)
10. Teahan, O., et al.: Impact of analytical bias in metabonomic studies of human blood serum and plasma. Anal. Chem. **78**(13), 4307–4318 (2006)
11. Gerdes, J.C. et al.: Kit Comprising Solid Phase Matrix Made of Aluminum Oxide, Titanium Oxide (Ti_2O_3) and/or Modified Zirconium Dioxide (ZrO_2) for Identifying and Extracting Viral Nucleotide Sequences. Google Patents (2006)
12. Wei, W., et al.: Elimination of the interference from nitrate ions on oxalic acid in RP-HPLC by solid-phase extraction with nanosized hydroxyapatite. J. Liq. Chromatogr. Relat. Technol. **32**(1), 106–124 (2008)
13. Monopoli, M.P., et al.: Physical—chemical aspects of protein corona: relevance to in vitro and in vivo biological impacts of nanoparticles. J. Am. Chem. Soc. **133**(8), 2525–2534 (2011)
14. Walczyk, D., et al.: What the cell "sees" in bionanoscience. J. Am. Chem. Soc. **132**(16), 5761–5768 (2010)
15. Lynch, I., Dawson, K.A.: Protein-nanoparticle interactions. Nano Today **3**, 40–47 (2008)
16. Salvati, A.: Transferrin-Functionalized Nanoparticles Lose Their Targeting Capabilities When a Biomolecule Corona Absorbs on the Surface. Nature Nanotech
17. Bajaj, A., et al.: Stability, toxicity and differential cellular uptake of protein passivated-Fe_3O_4 nanoparticles. J. Mater. Chem. **19**, 6328–6331 (2009)
18. Boraschi, D., Costantino, L., Italiani, P.: Interaction of nanoparticles with immunocompetent cells: nanosafety considerations. Nanomedicine **7**, 121–131 (2011)
19. Brown, D.M., et al.: Interaction between nanoparticles and cytokine proteins: impact on protein and particle functionality. Nanotechnology **21**, 215104 (2010)
20. Fadeel, B., Garcia-Bennett, A.E.: Better safe than sorry: understanding the toxicological properties of inorganic nanoparticles manufactured for biomedical applications. Adv. Drug Deliv. Rev. **62**, 362–374 (2010)
21. Ang, J.C., et al.: Protein trapping of silica nanoparticles. Soft Matter **6**, 383–390 (2010)
22. Brewer, S.H., et al.: Probing BSA binding to citrate-coated gold nanoparticles and surfaces. Langmuir **21**, 9303–9307 (2005)
23. Dobrovolskaia, M.A.: Interaction of colloidal gold nanoparticles with human blood: effects on particle size and analysis of plasma protein binding profiles. Nanomed. Nanotech. Biol. Med. **5**, 106–117 (2009)
24. Gagner, J.E., et al.: Effect of gold nanoparticle morphology on adsorbed protein structure and function. Biomaterials **32**, 7241–7252 (2011)
25. Schleh, C.: Size and surface charge of gold nanoparticles determine absorption across intestinal barriers and accumulation in secondary target organs after oral administration. Nanotoxicology **6**, 36–46 (2012)
26. Gasser, M.: The adsorption of biomolecules to multi-walled carbon nanotubes is influenced by both pulmonary surfactant lipids and surface chemistry. J. Nanobiotechnol. **8**, 31 (2010)
27. Ge, C.: Binding of blood proteins to carbon nanotubes reduces cytotoxicity. Proc. Natl. Acad. Sci. USA **108**, 16968–16973 (2011)
28. Docter, D., et al.: Quantitative profiling of the protein coronas that form around nanoparticles. Nat. Protocols **9**(9), 2030–2044 (2014)

Chapter 2
Methods

Abstract The nature of the research work that produced this thesis is highly interdisciplinary. Hence this chapter serves as a short primer on the techniques and methods that have been employed throughout this thesis. More detail has been provided for techniques which have been modified from their native forms.

2.1 Pulsed Laser Deposition

Pulsed laser deposition (PLD) is a highly controlled process to fabricate thin films with precise control over thickness and composition [1]. In a typical PLD setup, elementary or alloy targets are struck at an angle of 45° by a pulsed and highly focused laser beam. The atoms and ionic species ablated from the target(s) are deposited on a substrate placed parallel to the target surface at a target-to-substrate distance of typically 2–10 cm. A schematic of a simplified setup is shown in Fig. 2.1.

Since the energy source (laser) is outside the chamber, it is possible to maintain the chamber at ultra-high vacuum ($<10^{-6}$ Torr) or backfill it with a gas of choice (oxygen, nitrogen, argon). This control over the gaseous environment enables PLD to be used to fabricate thin films of a large class of compounds—oxides, nitrides, carbides. PLD has also been used for metallic [2] and polymer systems [3] in addition to Buckminister fullerenes [4].

But versatility is not PLD's only forte. The ability to reproduce the stoichiometry of the target (also referred to as "stoichiometry transfer") in the thin film is what made PLD popular. The superfast and intense heating of the target surface by the highly energetic laser beam (typically up to temperatures of more than 5000 K within a few nanoseconds, corresponding to a heating rate of about 10^{12} K/s) ensures that all target components, irrespective of their partial binding energies, evaporate at the same time [5]. When the ablation rate is sufficiently high (which normally is the case at laser fluences well above the ablation threshold), a so-called Knudsen layer is formed [6] and further heated (for instance by Inverse Bremsstrahlung) forming a high-temperature plasma [7], which then adiabatically

A. Patra, *Quantifying Interactions of Biomolecules with Inorganic Surfaces*, Springer Theses, DOI 10.1007/978-3-319-30728-2_2

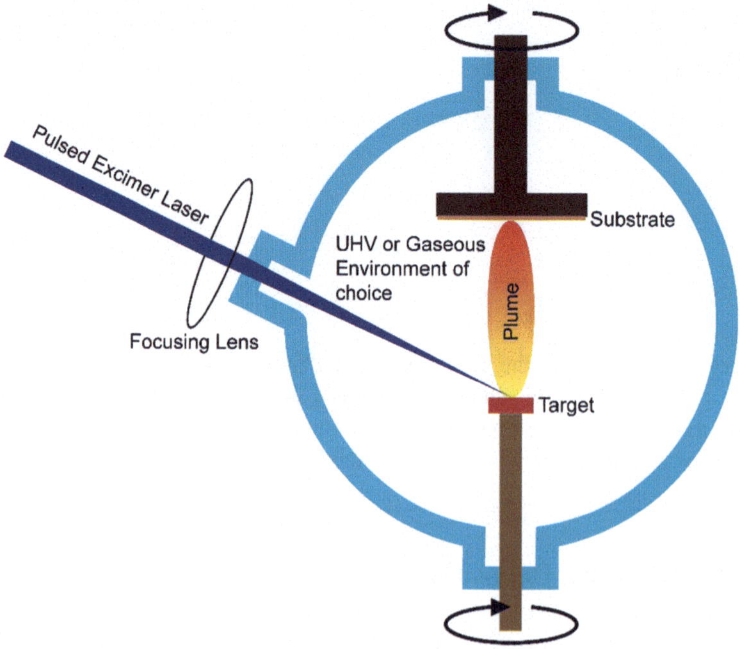

Fig. 2.1 A schematic layout of a simple PLD setup

swells in a direction perpendicular to the target surface. Therefore, during PLD, the material transfer between target and substrate occurs in a material package, where the separation of the species is small. Essentially, all complex structures like oxides and perovskites are built again at the substrate surface because all required components are transferred to the surface at almost the same time.

Furthermore, the laser fluence can be very easily adjusted. Polymers need a low 60 mJ/cm^2 to maintain stoichiometry in the thin film phase. However, metallic thin films need about 5 J/cm^2. It is possible to fabricate composites and heterostructures within a single PLD setup.

By altering laser fluence on the target and gas pressure within the chamber, the kinetic energy of the ionic species leaving the target within the plume can be altered which can lead to a lot of interesting changes in texture and microstructure.

In this work, PLD has been used as the deposition technique for fabricating high quality thin films throughout Chaps. 3 and 4.

2.2 Atomic Force Microscopy

The atomic force microscope (AFM) is perhaps the most common kind of scanning probe microscopes (SPM). SPMs are designed to measure local properties, such as topography, height, friction, magnetism, with a probe.

AFM [8] works by scanning a very sharp (end radius ca. 10 nm) probe along the sample surface, precisely sustaining the force between the probe and surface at a user-set, low level. Usually, the probe is made of a silicon or silicon nitride cantilever with a sharp integrated tip. The forces acting on the tip when it interacts with a sample causes bending of the cantilever. This bending or deflection is picked up by a laser focused on the back of the cantilever. The laser is reflected by the cantilever onto a distant position-sensitive photodetector. This system magnifies the normal bending of the cantilever greatly, and is sensitive to Angstrom-level movements. This set-up is also known as an optical lever. The probe is moved over the sample by a piezoelectric scanner, which can make extremely precise movements. The amount by which the scanner has to move in the z axis to maintain the cantilever deflection is taken to be equivalent to the sample topography. The combination of the sharp tip, the very sensitive optical lever, and the highly precise movements by the scanner, combined with the careful control of probe-sample forces allow the extremely high resolution of AFM.

AFM has been used to obtain images of the fabricated oxide surfaces and also to acquire fore-distance curves studied all through Chap. 3 (Fig. 2.2).

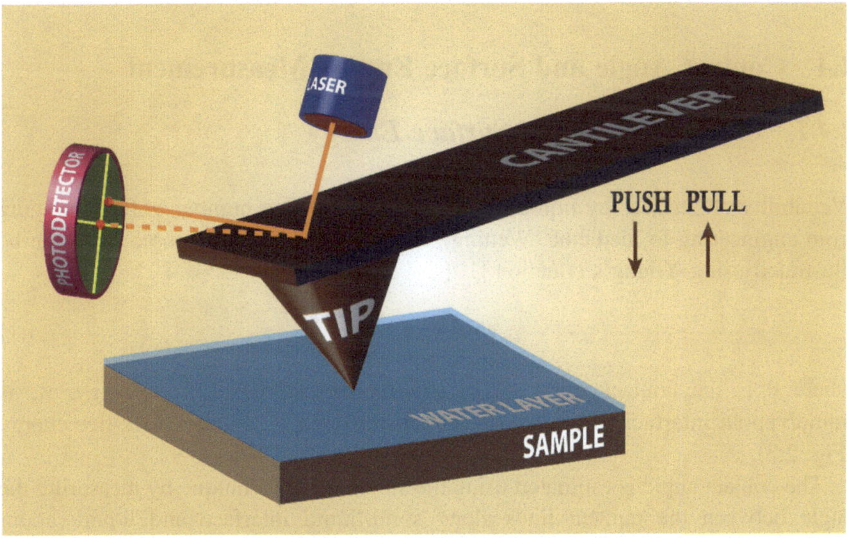

Fig. 2.2 Schematic of an AFM setup

2.3 Scanning Electron Microscopy

Scanning electron microscopes (SEM) use an accelerated beam of electrons to generate a variety of signals which are then collected by their respective detectors. Accelerated electrons in an SEM carry significant amounts of kinetic energy. Electron sample interactions cause deceleration of the incident electrons. The energy is then lost through various processes which generate a set of signals rich in information about the sample. These signals include secondary electrons (that produce SEM images), backscattered electrons (BSE), diffracted backscattered electrons (EBSD that are used to determine crystal structures and orientations of minerals), photons (characteristic X-rays that are used for elemental analysis and continuum X-rays), visible light (cathodoluminescence–CL), and heat [9]. While secondary electrons are most useful for generating morphological and topographical images of samples, backscattered electrons are used for acquiring contrasts in composition in multiphase or multicomponent samples. X-rays are generated by inelastic collisions of the incident electrons with electrons in discrete orbitals (shells) of atoms in the sample. When excited electrons in the sample return to lower energy states, they yield X-rays that possess a fixed wavelength (related to the difference in energy levels of electrons in different shells for a given element). Thus, characteristic X-rays are produced for each element in a sample that is hit by the incident electron beam.

SEM analysis is considered to be non-destructive. Electron interactions do not lead to volume loss of the sample, so it is possible to analyze the same samples repeatedly. SEM has been used in this work to find the particle size in metal oxide powders.

2.4 Contact Angle and Surface Energy Measurement

2.4.1 Contact Angle and Surface Energy

Wettability of surfaces by liquids is of great interest in a number of fields ranging from engineering to medicine. Wetting phenomena on a macroscopic scale can be illustrated using Young's equation [10]

$$\gamma_{lv}\cos\theta = \gamma_{sv} - \gamma_{sl} \quad (2.4.1)$$

where θ is the contact angle, γ_{sl} is solid/liquid interfacial free energy, γ_{lv} is liquid/vapour interfacial tension (surface tension) and γ_{sv} solid surface free energy (Fig. 2.3).

The contact angle is estimated using the sessile drop technique by measuring the angle between the tangent lines along solid-liquid interface and liquid-vapour interface of the liquid contour as shown in Fig. 2.3. A contact angle of 0° and 180°

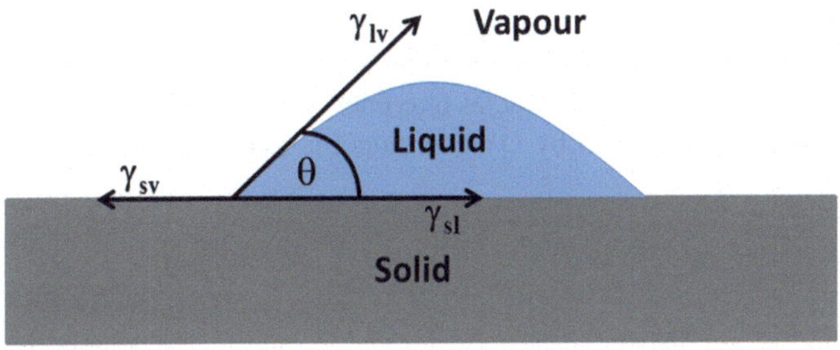

Fig. 2.3 Schematic diagram of a liquid drop on a solid surface showing the interfacial tensions at three phase boundary

correspond to complete wetting and non-wetting respectively. Surfaces exhibiting contact angles below 90° are called hydrophilic and those above 90° are called hydrophobic. In the past few decades many semi-empirical analytical models have been developed to compute surface free energy from measured contact angles such as Fowkes [11], Owens Wendt Rabel and Kaelbel (OWRK) [12], Van Oss-Chaudhury Good/Lewis acid base theory [13], Zisman [14] and Neumann [15] to name a few. Each approach is targeted for measuring surface energies of either low surface energy materials or high surface materials or both. Also, the essence and physical interpretation of these approaches are different and therefore subsequently provide information on total surface energy or individual components (polar, dispersive, hydrogen bonding etc.) of surface energy or both. In our work, wide range of samples ranging from hydrophilic to hydrophobic have been presented and therefore the OWRK method has been adopted which is suitable for universal systems.

2.4.2 Description of OWRK Method

Owens, Wendt, Rabel and Kaelble developed a two component model to separate the interfacial tension according to the underlying interactions between the molecules. These interactions are defined as polar and dispersive interactions. The total surface energy of the solid is the sum of the two parts. The polar interactions arise due to the permanent dipole—permanent dipole interactions or Keesom forces [16]. They are stronger and only exist in polar molecules. Dispersive component also known as London forces are weak and arise due to random fluctuations in the electron density in an electron cloud and hence lead to temporary/induced dipole interactions [17].

In OWRK method, at least two liquids with known dispersive and polar parts of surface tensions are needed to compute the solid surface free energy as there are two unknowns (solid/liquid interfacial free energy and solid surface free energy). The combining rule proposed by OWRK model is indicated below.

$$\gamma_{sl} = \gamma_{sv} + \gamma_{lv} - 2\left(\sqrt{\gamma_{sv}^{D}\gamma_{lv}^{D}} + \sqrt{\gamma_{sv}^{P}\gamma_{lv}^{P}}\right) \tag{2.4.2}$$

where γ_{sv}^{D} and γ_{lv}^{D} are dispersive components and γ_{sv}^{P} and γ_{lv}^{P} are polar components of solid and liquid surface energies respectively.

Substituting for γ_{sl} from Eq. (2.4.1),

$$\sqrt{\gamma_{sv}^{D}\gamma_{lv}^{D}} + \sqrt{\gamma_{sv}^{P}\gamma_{lv}^{P}} = \frac{1}{2}[\gamma_{sv} + \gamma_{lv} - (\gamma_{sv} - \gamma_{lv}\cos\theta] \tag{2.4.3}$$

$$\sqrt{\gamma_{sv}^{D}\gamma_{lv}^{D}} + \sqrt{\gamma_{sv}^{P}\gamma_{lv}^{P}} = \frac{1}{2}[\gamma_{lv}(1 + \cos\theta] \tag{2.4.4}$$

By dividing $\sqrt{\gamma_{lv}^{D}}$ in Eq. (2.4.4), we get,

$$\sqrt{\gamma_{sv}^{D}} + \sqrt{\gamma_{sv}^{P}}\sqrt{\frac{\gamma_{lv}^{P}}{\gamma_{lv}^{D}}} = \frac{1}{2}\frac{[\gamma_{lv}(1 + \cos\theta]}{\sqrt{\gamma_{lv}^{D}}} \tag{2.4.5}$$

The above equation can be represented in the linear form,

$$c + mx = y \tag{2.4.6}$$

where in, $c = \sqrt{\gamma_{sv}^{D}}$ $m = \sqrt{\gamma_{sv}^{P}}$ $x = \sqrt{\frac{\gamma_{lv}^{P}}{\gamma_{lv}^{D}}}$

A graphical representation of the OWRK method is shown in Fig. 2.4 for lubricated hard disk media sample. The polar [SFTP(l)] and dispersive components [SFTD(l)] of total surface tension [SFT(l)] of the liquids used in this study are known (Table 2.1) and are substituted to compute the polar and dispersive components of the surface free energy of the solid. The slope of the graph gives the polar component and the vertical intercept gives the dispersive component of the solid surface free energy.

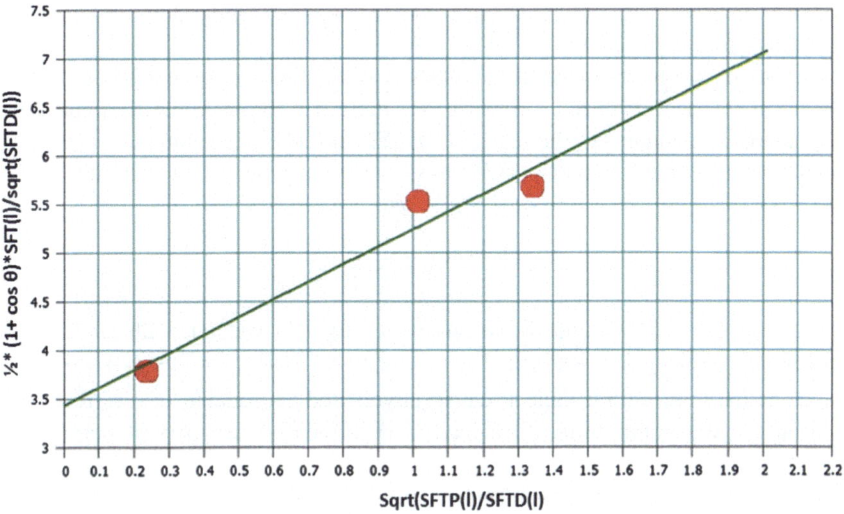

Fig. 2.4 Surface energy (polar and dispersive) calculation graph based on OWRK model for lubricated hard disk media

Table 2.1 Surface tensions of the test liquids

Label	SFT (total) mN/m	SFT (dispersive) mN/m	SFT (polar) mN/m	Values adapted from
Water	72.80	26	46.80	Gebhardt et al.
Ethylene Glycol	47.70	26.40	21.30	Gebhardt et al.
Diiodomethane	50.00	47.40	2.60	Busscher et al.

2.5 Surface Plasmon Resonance

2.5.1 Surface Plasmons

Surface plasmons (SPs), sometimes also called surface plasmon polaritons, are electromagnetic waves that propagate in a direction parallel to the metal/dielectric (or metal/vacuum) interface, along the surface of separation. In 1902, Wood [18] illuminated a metallic diffraction grating with polychromatic light and noticed narrow dark bands in the spectrum of the diffracted light. He almost dismissed them as unexplained anomalies. Fano followed up on Wood's report with theoretical work, diligently explaining these anomalies to be associated with the excitation of electromagnetic surface waves on the surface of the diffraction grating. Thereafter, Otto [19], Kretschmann and Raether, Agerwal [20], and Swalen [21] have performed exhaustive and comprehensive work which has not only successfully established an easy method but also a verifiable theory for the excitation of surface plasmons and their investigation, consequently opening up the domain of surface plasmons in modern optics.

Surface plasmons originate from collective oscillations of the quasi-free electron gas at the interface between a metal and a dielectric. From Maxwell's equations with standard boundary conditions, it is possible to show that an interface between a metal and a dielectric can support SP mode with the propagation constant β as:

$$\beta = k_0\sqrt{\frac{\varepsilon_d\varepsilon_m}{\varepsilon_d + \varepsilon_m}} \tag{2.5.1}$$

where $k_0 = \omega/c = 2\pi/\lambda$ is the wave number of light in vacuum with angular frequency ω, wavelength λ, and speed c. $\varepsilon_m = \varepsilon'_m + i\varepsilon''_m$ and ε_d are the permittivity of metal and the semi-infinite dielectric respectively and i is equal to $\sqrt{-1}$. Metals like gold, silver and aluminium which have a negative and real part of permittivity ($\varepsilon'_m < 0$) can support SPs.

Assuming $|\varepsilon'_m| \gg \varepsilon''_m$ the propagation constant β of the surface plasmon can be written as

$$\beta = Re\{\beta\} + iIm\{\beta\}$$

Plugging in the values from equation

$$\beta = \frac{\omega}{c}\sqrt{\frac{\varepsilon_d\varepsilon'_m}{\varepsilon_d + \varepsilon'_m}} + i\frac{\varepsilon'_m}{2(\varepsilon'_m)^2}\frac{\omega}{c}\left(\frac{\varepsilon_d\varepsilon'_m}{\varepsilon_d + \varepsilon'_m}\right)^{\frac{3}{2}} \tag{2.5.2}$$

where Re{ } and Im{ } denote the real and imaginary part of a complex number, respectively. The imaginary part of the propagation constant is associated with the attenuation of the surface plasmon propagation.

At the metal dielectric interface, the amplitude of the SP electromagnetic field is at its maximum. Away from the interface, it decays, exponentially into both the metal and dielectric media, as shown later. L_p is used to represent the field decay in the direction perpendicular to the metal dielectric interface. It is defined as the distance from the interface at which the amplitude of the field decreases by a factor of 1/e.

For the metal, the penetration depth expressed as

$$L_{pm} = 1/Im\{k_{zm}\} = 1/Im\left(\frac{\omega}{c}\frac{\varepsilon_m}{\sqrt{\varepsilon_m + \varepsilon_d}}\right) \tag{2.5.3}$$

For the dielectric, similarly

$$L_{pm} = 1/Im\{k_{zd}\} = 1/Im\left(\frac{\omega}{c}\frac{\varepsilon_d}{\sqrt{\varepsilon_m + \varepsilon_d}}\right) \tag{2.5.4}$$

Z being the direction into the metal and dielectric (going into plane of paper in Fig. 2.7) Propagation losses in the metal cause damping of surface plasmons travelling along a metal/dielectric interface. The finite propagation length L_x of SPs is defined as the distance in the x-direction where the energy of the surface plasmon has decreased by a factor of 1/e.

$$L_x = \frac{1}{2Im\{\beta\}} \tag{2.5.5}$$

2.5.2 *Excitation of Surface Plasmons*

Theoretically, surface plasmons can be excited by the bombarding the metal with electrons. However, due to the ease of experimental implementation optical excitation is more commonly used. The SP dispersion curve asymptotically approaches the light line at low energies as in Fig. 2.5 whereas for higher energies it approaches the cutoff angular frequency ω_{max} that is proportional to the metal plasma frequency

$$\omega_{\max} = \frac{\omega_p}{\sqrt{1+\varepsilon_d}} \tag{2.5.6}$$

Considering that the momentum of free photons travelling through a dielectric medium (2.5a) is always smaller than the propagation constant Re{β} of SPs (2.5c) at the dielectric/metal interface

$$k_{ph} = \frac{\omega}{c}\sqrt{\varepsilon_d} < Re\left\{\sqrt{\frac{\varepsilon_d\varepsilon_m}{\varepsilon_d+\varepsilon_m}}\right\} = Re\{\beta\} \tag{2.5.7}$$

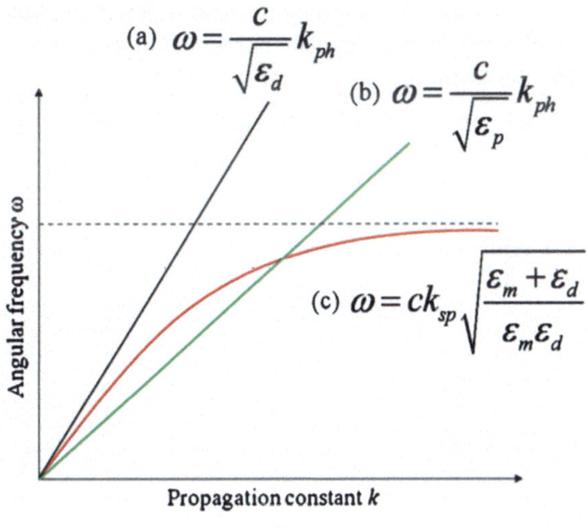

Fig. 2.5 The dispersion relation of free photons in a dielectric (**a**), and in a coupling prism (**b**) with $\varepsilon_p > \varepsilon_d$, compared to the dispersion relation of non-radiative surface plasmons (**c**) at the metal/dielectric interface

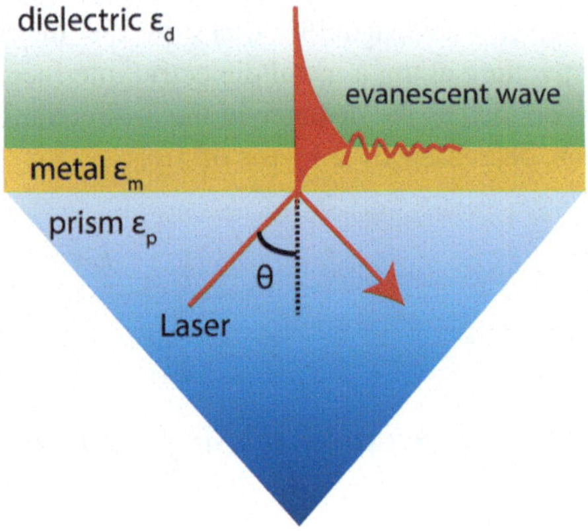

Fig. 2.6 A schematic of how SPs are excited in the Kretschmann configuration

These two waves cannot be phase matched along the metal surface due to the insufficient k_{ph}. Therefore they cannot be directly coupled. To optically excite SPs, we have to employ a coupling enlargement. In this work, we use a prism coupler and use the attenuated total reflection method (ATR). In the Kretschmann geometry (used throughout Chap. 4) a laser beam is incident on a 50 nm gold film through a prism of permittivity ε_p. The scheme is illustrated in Fig. 2.6. If $Re\{\beta\}$ of surface plasmon at the interface between the metal and the lower refractive index dielectric is matched to the one of the incident optical wave, surface plasmons are generated

$$\frac{\omega}{c}\sqrt{\varepsilon_p}\sin\theta = Re\{\beta\} \tag{2.5.8}$$

The strength of coupling between the incoming optical wave (laser) and the surface plasmons wave, therein excited, can be changed by altering the thickness of the metal layer. Excitation of SPs manifests itself as a narrow absorption dip in angular spectrum reflectivity as measured with a photodetector placed in the path of the reflected laser beam. The particular angle at this which occurs can be calculated from the formula

$$\theta = \arcsin Re\left\{\sqrt{\frac{\varepsilon_d\varepsilon_m}{(\varepsilon_d+\varepsilon_m)\varepsilon_p}}\right\} \tag{2.5.9}$$

A typical angular spectrum of reflectivity is shown in Fig. 2.7.

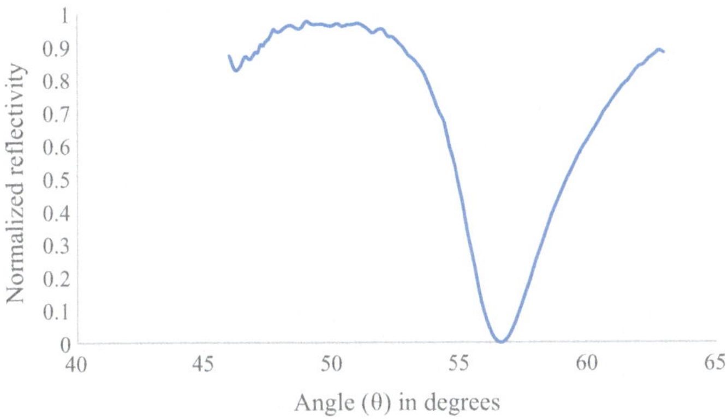

Fig. 2.7 Angular spectrum of reflectivity showing a dip at 56°

2.5.3 Optical Sensing with Surface Plasmon Resonance

In the beginning, surface plasmons were used in the characterization of thin films and the study of processes at metal boundaries. Since then, however, surface plasmons have marked their entry in non-linear optics, lasers, optical data storage, photonic bandgap structures, sub-diffraction imaging and optical sensors. We will now discuss how SPR has been used for biosensing purposes.

2.5.4 Working Principle of SPR Biosensors

If we take the most reductionist view, SPR biosensors are thin-film refractometers that have the ability to measure extremely tiny changes in the refractive index occurring at the surface of a metallic films which support the excitation of SPs. SPR biosensors consist of two major components

(a) The optical setup for the excitation and interrogation of SPs
(b) Biomolecular recognition elements (BRE)

The BRE are typically immobilized on the sensor surface or in the sensitive layer (e.g. alginate matrix in Chap. 5 of this study) probed by evanescent field of resonantly excited SPs. Whenever analyte molecule binds to BRE, the refractive index in the close proximity of the gold (or SP supporting metallic surface) changes by δn, which in turn causes a change of propagation constant δβ. The δβ changes the coupling condition to SPs, as indicated in Eqs. (2.5.1) and (2.5.8). So by measuring variations of the characteristics of the light wave coupled to the SP wave at the metal dielectric interface, we can investigate molecular binding events with the BRE.

In Chap. 4, the BRE we are using is a thin layer of transition metal oxide which have certain affinity towards phospholipids. This is a departure from the normal use

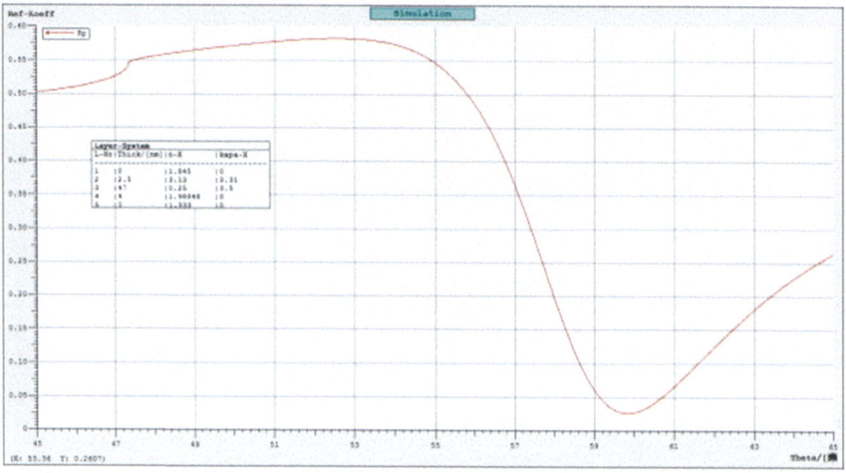

Fig. 2.8 Simulation of angular spectrum of reflectivity of SC_2O_3 (5 nm) film on a gold (50 nm) film in water. Resonance occurs at 59.9°

scenario where sensing elements are covalently bonded to the gold surface. When dealing with a thin film of metal oxide on the surface of gold, there are two issues which need to be tackled.

1. **Thickness**—The thickness of the film cannot be more than 3 nm. This is because at higher thicknesses, the angle at which resonance is observed becomes greater than 60°. The setup could not measure reflectivity at those angles because of geometrical constraints. It must be noted, most commercial setups will not measure binding events at such angles as well. Hence, by using equations shown above, we could simulate the angle which could be measured within our limitations and get to know what thickness of films will work for us. A simulation for 5 nm film of SC_2O_3 is shown in Fig. 2.8.
2. **Adhesion of oxide film to gold**—Usually, sensing elements are covalently bonded to the gold surface through some chemical linker molecules. These tend to be very stable on the surface. In the case of oxide thin films, this issue is non-trivial. We use the PLD process described in Sect. 2.1 to deposit the thin film onto the gold surface. However, as we learned through experiment, it is difficult to stabilize thin films grown at room temperature on the surface. Under flow of buffer solution, these films would simply delaminate and get washed away. We even had cases of chunks of films breaking off, which could be monitored by the SPR response. Usually, in the PLD process, by raising deposition temperatures above 500 °C, adhesion to substrate is improved. But, the gold would boil off under the high vacuum and 500 °C conditions inside the chamber. Through rigorous trials, we could optimize the deposition conditions under which we could obtain highly stable metal oxide thin films on gold. Thereafter, we are able to monitor very precisely the amount of phospholipids which bind to the thin film of transition metal oxide.

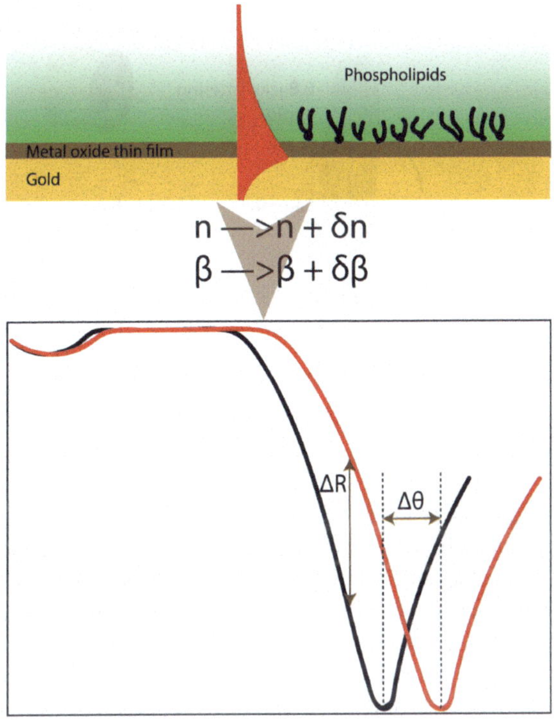

Fig. 2.9 Working principle of SPR biosensor in angular mode of operation, used in Chap. 4 to find amount of phospholipid bound to thin film of metal oxide

In SPR biosensor with angular modulation (used in Chap. 4), a monochromatic light wave (632.8 nm laser in this case) is used to excite an SP wave. There are two parameters that can be monitored to record molecular binding events taking place on the surface. One is the angular SPR reflectivity dip $\Delta\theta$ and the other is reflectivity change ΔR at an angle set very near to the resonance dip. The principle of operation of SPR biosensors measuring the binding-induced refractive index changes as a shift in the SPR resonant angle $\Delta\theta$, or the changes of reflectivity ΔR is shown in Fig. 2.9.

2.6 Rutherford Backscattering Spectroscopy

Rutherford Backscattering Spectroscopy (RBS) is named after Ernest Rutherford who discovered the atomic nucleus in the very famous experiment which involved the bombarding of a very thin gold foil by energetic alpha particles (He^{2+}) in the year 1911. Rubin et al. [22] were the first scientists to use nuclear scattering for detection of elements up to a depth of several μm. The method gained popularity with the invention of solid state detectors in 1970. The interest fueled a lot of scientific efforts and investment which consequently made RBS a very powerful

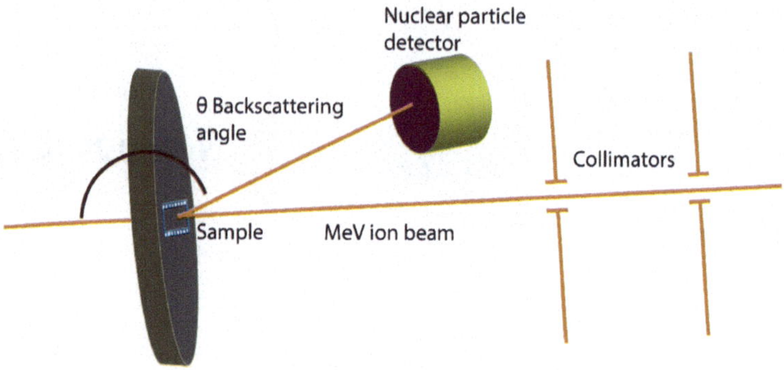

Fig. 2.10 Rutherford backscattering spectroscopy layout

technique for structural and compositional characterization of thin films or single crystals.

An ion beam of energetic (typically few MeV) beam of a (He^{2+}) particles is produced by an electrostatic accelerator and is directed towards the sample by the use of magnets. Some part of the incident beam is backscattered due to collisions with nucleus of elements present in the sample. These collisions can be accurately modeled as an elastic collision using classical physics. The energy of the backscattered ion is a function of atomic mass of the element and it asymptotically approaches the energy of the incident particles as the mass of target atom increases. The number and energy of the backscattered ions from the sample is detected. The energy of the backscattered ions is used to identify elements and the count is used to get elemental concentration [23].

Also energy of the ions which are backscattered by the same element at some depth in the film is lower than the one on the surface. This energy loss is dependent on sample composition and density. This is used for measurement of film thickness and depth profiling. The count of backscattered alpha particles as a function of energy gives a spectrum which is analyzed to get the chemical composition and thickness of the film.

In this thesis, RBS has been used only for checking elemental composition of thin films and obtaining their thickness (Fig. 2.10).

2.7 Mass Spectrometry

John B. Fenn, the inventor of electrospray ionization for biomolecules and the 2002 Nobel Laureate in Chemistry, described mass spectrometry as follows.

"Mass spectrometry is the art of measuring atoms and molecules to determine their molecular weight. Such mass or weight information is sometimes sufficient, frequently necessary, and always useful in determining the identity of a species.

To practice this art one puts charge on the molecules of interest, i.e., the analyte, then measures how the trajectories of the resulting ions respond in vacuum to various combinations of electric and magnetic fields.

Clearly, the sine qua non of such a method is the conversion of neutral analyte molecules into ions. For small and simple species the ionization is readily carried by gas-phase encounters between the neutral molecules and electrons, photons, or other ions. In recent years, the efforts of many investigators have led to new techniques for producing ions of species too large and complex to be vaporized without substantial, even catastrophic, decomposition."

There are four major components of a mass spectrometer

A. Sample inlet—to introduce the sample in a compatible format into the mass spectrometer
B. Ionization source—to generate ions from the sample
C. Mass analyzer—to sort ions according to mass and charge
D. Ion detector—to measure the separated ions.

Two types of ionization sources have been used in this thesis, hence we would enter a short discussion about it.

2.7.1 Matrix Assisted Laser Desorption Ionization

Tanaka, Karas, and Hillenkamp introduced matrix-assisted laser desorption/ ionization mass spectrometry (MALDI-MS) in 1988 [24, 25]. MALDI is a soft ionization technique in which the energy from the laser is used to volatilize the matrix rather than degrading the sample. This technique is based upon an ultraviolet absorbing weak organic acid. The matrix and sample are dissolved in an appropriate solvent which also prevents aggregation of the sample. The sample–matrix mixture is placed under vacuum conditions; the solvent is removed, leaving behind co-crystallized sample molecules hopefully, homogenously dispersed within matrix molecules. The exact mechanism is not known but it is believed that when the pulsed UV laser is tuned to the appropriate frequency, it causes the ionization and transfer of a sample from the condensed phase to the gas phase via laser excitation and ablation of the sample. The co-crystallized sample molecules vaporize instantly without having to directly absorb energy from the laser. The directed energy transfer during a matrix-assisted laser-induced desorption event provides quite high ion yields of the intact sample, and allows for the measurement of compounds with sub-picomole sensitivity.

Once vaporized, the desorbed charged molecules are guided by an electric field from the MALDI ionization source into the mass analyzer. Most commonly, time-of-flight (TOF) mass analyzers are used to sort the ions according to their mass-to-charge ratio (m/z).

The popularity of MALDI for biomolecule analyses is due to its high sensitivity, suitability for application to complex mixtures, tolerance for salts in millimolar concentrations in addition to the fact that it is a soft ionization method which causes little or no fragmentation of species.

In Sect. 4.3.1, we have deposited thin films of metal oxides on the commercially obtained MALDI target plate for the purpose of testing efficacy of phospholipid capture by these thin films.

2.7.2 Electrospray Ionization

Dole carried out the practical development of electrospray ionization (ESI) though the first experiments were performed by Chapman in the 1930s. Subsequently, he also discovered the important phenomenon of multiple charging of molecules. However, it was truly Fenn's work [26] (for which he was awarded the Nobel Prize) that ultimately led to the modern day technique of electrospray ionization mass spectrometry and its application to biological macromolecules.

Within the ESI setup, the voltage applied (700–5000 V) to the needle produces an electrical gradient on the fluid flowing through. Either dry gas, heat, or both are applied to the droplets at atmospheric pressure thus causing the solvent to evaporate from each droplet. This electrical gradient separates the charges at the surface of the flowing fluid. Consequently, the fluid is forced to emerge from the needle as a Taylor cone. The tip of the Taylor cone, thus formed, keeps protruding as s fine filament until such a point where the fluid reaches the Rayleigh limit where the surface tension of the fluid equals the electrostatic repulsion. At this point, highly charged droplets leave the filament. The charged droplets that have emanated from the filament are attracted to the entrance of the mass spectrometer (through electrostatic lenses) due to the high opposite voltage at the mass analyzer's entrance. As these droplets now traverse the distance towards the mass analyzers, the Coulombic repulsion generated by the similar charges on the droplets' surface comes to exceed the surface tension holding the droplets together. When this happens, the droplets burst into much smaller droplets thereby also releasing the ions. Hence, the overall function of ESI is to produce gaseous ionized molecules directly from a liquid solution.

Electrospray ionization is regularly used for analyzing peptides, proteins, carbohydrates, small oligonucleotides, synthetic polymers, and lipids. Since ESI involves the continuous flow of solution, it is highly suitable for using as an interface with High Performance Liquid Chromatography (HPLC) or capillary electrophoresis.

NanoESI has become popular in recent years. This setup is a variant of the standard ESI, whereby advances in instrumentation has made it possible to have an extremely small spray needle which is placed exceedingly close to the mass analyzer. Such an arrangement ultimately results in an increase in efficiency which essentially means that smaller amounts of sample are required.

In this thesis, ESI-MS has been used to evaluate lipid fraction enrichment through metal oxide μcolumns.

2.8 Liquid Chromatography

The earliest reports, perhaps pioneering, of liquid chromatography as we know it today are the work of Russian botanist, Tswett [27]. The flash of brilliance he had was to separate leaf pigments (extracted from leaves using an organic solvent) by running them through a column packed with chalk (calcium carbonate) and alumina particles. He followed this with a run of pure solvent. As the force of gravity pulled the sample through the column, the various pigments separated into differently colored bands. Tsewett had thereby, succeeded to analytically separate different components in a complex mixture exploiting their inherent distinct partitioning ratio between the moving solvent (mobile phase) and the chalk or alumina particles (stationary phase).

We will take a look at two subtypes of liquid chromatography that has been used in this work.

2.8.1 Thin Layer Chromatography

Thin layer chromatography (TLC) is a planar version of column chromatography. The stationary phase is, in this case, a thin layer of silica or alumina, immobilized onto a metal or glass plate. The sample to be analyzed is spotted using a micropipette near the base of the plate. A volatile organic solvent is filled in the container as shown in the figure. The solvent level is so adjusted such that is just touches the spot. The container is then sealed of so that the environment inside the chamber is saturated with vapor of the organic solvent. The solvent carries the sample as it moves by capillary action through the dry particle bed thereby partitioning the various components according to their relative affinities for the stationary phase and the mobile phase.

2.8.2 High Performance Liquid Chromatography

As its name aptly suggests, high performance liquid chromatography (HPLC) is a technological upgrade on liquid chromatography. Instead of using gravity as the driving force for dragging the sample through the column, HPLC uses a powerful pump which pushes the sample through the column at pressures exceeding 6000 psi (400 bar).

The acronym HPLC was coined by late Prof. Csaba Horváth in 1970 when pumps had a maximum capability of 500 psi or 35 bar. He called it high pressure liquid chromatography [28]. However, advances in technology and instrumentation soon made it possible for pumps to generate 6000 psi. When this advancement was coupled with more advanced detectors, sample injectors and better columns, HPLC really began to establish itself as a very powerful tool in analytical chemistry. Over time, the expansion of the acronym was changed to high performance liquid chromatography.

Today a lot of scientific effort is directed towards making very sensitive and durable columns which make it easy to separate compounds present even in parts per trillion concentrations. Due to these capabilities, HPLC has become indispensable to proteomics, lipidomics, pharmaceuticals, food, nutraceuticals, cosmetics, environmental matrices, forensic samples, and industrial chemicals domains.

2.8.2.1 Normal Phase HPLC

In his pioneering experiment, Tsewett had used a polar stationary phase (chalk and alumina) and a much less polar mobile phase to accomplish the segregation of the dyes in the leaf extract. This scheme came to be generically known as normal phase HPLC, where the stationary phase is much more polar than the mobile phase. Hence, due to polar-polar interactions, the polar components in the complex mixtures are retained much more strongly than the apolar components.

2.8.2.2 Reverse Phase HPLC

As is obvious from the nomenclature, this is the scheme where the conditions of the mobile and stationary phase are reverse of the normal phase. Hence, in reverse phase HPLC (RP-HPLC), the mobile phase is much more polar than the stationary phase. Consequently, apolar components are retained most strongly and polar components are eluted quicker.

References

1. Dijkkamp, D., et al.: Preparation of Y-Ba-Cu oxide superconductor thin films using pulsed laser evaporation from high Tc bulk material. Appl. Phys. Lett. **51**(8), 619–621 (1987)
2. Krebs, H.U., Bremert, O.: Pulsed laser deposition of thin metallic alloys. Appl. Phys. Lett. **62** (19), 2341–2343 (1993)
3. Hansen, S., Robitaille, T.: Formation of polymer films by pulsed laser evaporation. Appl. Phys. Lett. **52**(1), 81–83 (1988)
4. Curl, R.F., Smalley, R.E.: Fullerenes. Sci. Am. **265**, 54–63 (1991)

5. Fähler, S., Krebs, H.-U.: Calculations and experiments of material removal and kinetic energy during pulsed laser ablation of metals. Appl. Surf. Sci. **96**, 61–65 (1996)
6. Kelly, R., et al.: Laser sputtering: Part I. On the existence of rapid laser sputtering at 193 nm. Nucl. Instrum. Methods Phys. Res., Sect. B **9**(3), 329–340 (1985)
7. Phipps Jr., C., et al.: Impulse coupling to targets in vacuum by KrF, HF, and CO_2 single-pulse lasers. J. Appl. Phys. **64**(3), 1083–1096 (1988)
8. Binnig, G., Quate, C.F., Gerber, C.: Atomic force microscope. Phys. Rev. Lett. **56**(9), 930 (1986)
9. Goldstein, J., et al.: Scanning Electron Microscopy and X-ray Microanalysis: A Text for Biologists, Materials Scientists, and Geologists. Springer Science & Business Media, Berlin (2012)
10. Young, T.: An essay on the cohesion of fluids. Philos. Trans. R. Soc. Lond 65–87 (1805)
11. Fowkes, F.M.: Calculation of work of adhesion by pair potential summation. J. Colloid Interface Sci. **28**(3–4), 493–505 (1968)
12. Owens, D.K., Wendt, R.C.: Estimation of the surface free energy of polymers. J. Appl. Polym. Sci. **13**(8), 1741–1747 (1969)
13. Van Oss, C.J., Good, R.J., Chaudhury, M.K.: The role of van der Waals forces and hydrogen bonds in "hydrophobic interactions" between biopolymers and low energy surfaces. J. Colloid Interface Sci. **111**(2), 378–390 (1986)
14. Fox, H., Zisman, W.: The spreading of liquids on low-energy surfaces. II. Modified tetrafluoroethylene polymers. J. Colloid Sci. **7**(2), 109–121 (1952)
15. Zenkiewicz, M.: New method of analysis of the surface free energy of polymeric materials calculated with Owens-Wendt and Neumann methods. Polimery **51**(7–8), 584–587 (2006)
16. Keesom, W.: The second viral coefficient for rigid spherical molecules, whose mutual attraction is equivalent to that of a quadruplet placed at their centre. In: Proceedings of the Royal Netherlands Academy of Arts and Sciences (KNAW), Amsterdam (1915)
17. London, F.: The general theory of molecular forces. Trans. Faraday Soc. **33**, 8b–26 (1937)
18. Wood, R.W.: On a remarkable case of uneven distribution of light in a diffraction grating spectrum (1902)
19. Otto, A.: Excitation of nonradiative surface plasma waves in silver by the method of frustrated total reflection. Zeitschrift für Physik **216**(4), 398–410 (1968)
20. Agarwal, G.: New method in the theory of surface polaritons. Phys. Rev. B **8**(10), 4768 (1973)
21. Swalen, J.: Optical properties of Langmuir-Blodgett films. J. Mol. Electron. **2**, 155–181 (1986)
22. Rubin, S.: Surface analysis by charged particle spectroscopy. Nucl. Instrum. Methods **5**(3), 177–183 (1959)
23. Chu, W.-K., Backscattering spectrometry. Elsevier (2012)
24. Karas, M., Hillenkamp, F.: Laser desorption ionization of proteins with molecular masses exceeding 10,000 daltons. Anal. Chem. **60**(20), 2299–2301 (1988)
25. Karas, M., et al.: Matrix-assisted ultraviolet laser desorption of non-volatile compounds. Int. J. Mass Spectrom. Ion Process. **78**, 53–68 (1987)
26. Yamashita, M., Fenn, J.B.: Electrospray ion source. Another variation on the free-jet theme. J. Phys. Chem. **88**(20), 4451–4459 (1984)
27. Tswett, M.: Physikalisch-chemische Studien über das chlorophyll. Die Adsorptionen. Berichte der Deutschen botanischen Gesellschaft **24**(316–323): 20 (1906)
28. Horvath, C., Lipsky, S.: Column design in high pressure liquid chromatography. J. Chromatogr. Sci. **7**(2), 109–116 (1969)

Chapter 3
Quantifying Interactions Between Water and Surfaces

Abstract In this chapter, we deploy an atomic force microscopy-based approach to locally probe surface wettability. The maximum pull-off force as calculated from force spectroscopy shows a remarkable correlation with the macroscale water contact angle, measured over a wide variety of surfaces starting from hydrophilic, all the way through to hydrophobic ones. This relationship, consequently, facilitates the establishment of a universal behavior. The adhesion and capillary forces scale with the polar component of surface energy. However, no such relation could be established with the dispersive component. Hence, we postulate that the force(s) which enable us to correlate the force spectroscopy data measured on the nano scale to the macroscopic contact angle are primarily arising from electrostatic-dipole-dipole interactions on the surface. London forces play less of a role. This effect in is line with density functional theory (DFT) calculations suggesting a higher degree of hydroxylation of hydrophilic surfaces. This result shows that molecular simulations and measurements on an atomic scale can be extrapolated to macroscopic surface wetting problems.

3.1 Introduction

3.1.1 Overview

In the context of the organization of this thesis work, we begin our studies by trying to understand how solid surfaces interact with perhaps the most common and indispensable liquid molecule—water. Solid-liquid interfaces have been of interest in photo-catalysis, water splitting and the design of novel super-hydrophilic/hydrophobic surfaces [1, 2]. The fundamental understanding of interfacial phenomenon has been crucial for various self-assembled monolayers (SAM) [3–5], biological processes [6–8] and technologies such as dip-pen lithography [9, 10]. Intrinsic wetting properties arising from inherent surface chemistry have attracted particular interest recently [2]. Wetting properties and surface free energies at macro-scale are ultimately governed by the intermolecular interactions and surfaces

A. Patra, *Quantifying Interactions of Biomolecules with Inorganic Surfaces*,
Springer Theses, DOI 10.1007/978-3-319-30728-2_3

forces on the nanoscale. Conventional methods such as contact angle measurements [11] and inverse gas chromatography [12] are employed widely for measuring these wetting properties.

3.1.2 Literature Review and the Need for a New Technique

With the development of nano and micro-scale devices e.g., fluidic devices wetting measurements on small dimensions become important. However, the above methods (like contact angle measurements [11] and inverse gas chromatography [12] techniques) do not work on the nanometer scale and there is a need for the development of alternative approaches. Some efforts have been made recently to understand the solid-liquid interface of a hydrophilic surface using nano-ultrasonics [13]. Several numerical simulations have been performed to understand the intriguing phenomena that happen at nanoscale and how they go on to affect the macroscale wetting properties as measured through contact angle [14, 15]. The nature of adhesion forces responsible for these phenomena has been a topic of debate [11]. A significant amount of work has been reported on the study of adhesion properties using atomic force microscopy (AFM) [11, 16–18]. At this point, it is important to note that the capillary forces between the AFM tip and the surfaces form a significant part of the overall adhesion forces [11, 16, 18, 19]. However, the capillary forces alone cannot explain the large adhesion observed on hydrophilic surfaces without assuming other strong liquid-solid interaction [20, 21]. Eastman et al reported that the adhesion force is significantly reduced when a hydrophobic tip is used [16]. The large reduction of adhesion cannot be explained by van der Waals interactions in Derjaguin-Muller-Toporov (DMT) and Johnson-Kendall-Roberts (JKR) methods. Hence, there is obviously another numerically dominant force, possibly originating from the intrinsic chemical nature of surfaces.

3.1.3 Motivation and Rationale

To plug existing gaps in our knowledge of interfacial phenomenon, this force needs to be quantified and related to other measurable surface properties. With this aim, we combine macroscale contact angle measurements with force spectroscopy (FS) to ultimately demonstrate that the maximum pull-off force (adhesion) is an exponential function of macroscale water contact angle and surface energy. We would like to point out that this force correlates well with the polar component of the surface energy while finding no such relation with the dispersive component. This is an unambiguous pointer to the nature of this force that we have measured. We can, therefore, say that the measured force arises from electrostatic-dipole-dipole interaction (hydrogen bonding) on the surface and London forces (induced-dipole-dipole interactions) do not play a significant role. Density functional theory (DFT) calculations indicate the tendency of hydrophilic surfaces to stabilize hydroxyl groups, providing the linkage to create interfacial hydrogen bonds with liquid water.

3.2 Results and Discussion

3.2.1 Measurement of Contact Angles on Surfaces and Calculation of Surface Energies

Figure 3.1 shows the contact angle measurement results on different surfaces. Figure 3.1a is a schematic of vapor-liquid-solid interfaces where Θ, λ_{lv}, λ_{sv}, λ_{sl}, are contact angle, interfacial free energies of liquid/vapor, solid/vapor, solid/liquid respectively. Figure 3.1b–i shows samples of water contact angle ranging from 0° (superhydrophilic) to ~102° (hydrophobic). The contact angle is related to interfacial energies through Young's equation $\lambda_{sv} = \lambda_{sl} + \lambda_{lv}\ cos\ (\Theta)$. The contact angles have an intrinsic component which originates from the specific chemical composition of the surface and unsaturated intramolecular bonds thereon, and an external component due to roughness or any periodic structures.

In order to probe the local forces arising solely due to the intrinsic chemical nature, we chose smooth surfaces of comparable root mean square (RMS) roughness measured by AFM as shown in Fig. 3.2. There is a small variation in the amount of roughness between the films as can be observed (Fig. 3.2) but it does not influence the intrinsic surface energy. It has been widely reported that roughness in any hydrophilic surface reduces the water contact angle [11]. However, Sun et al have shown that surface features below ~50 nm will not trap sufficient air to

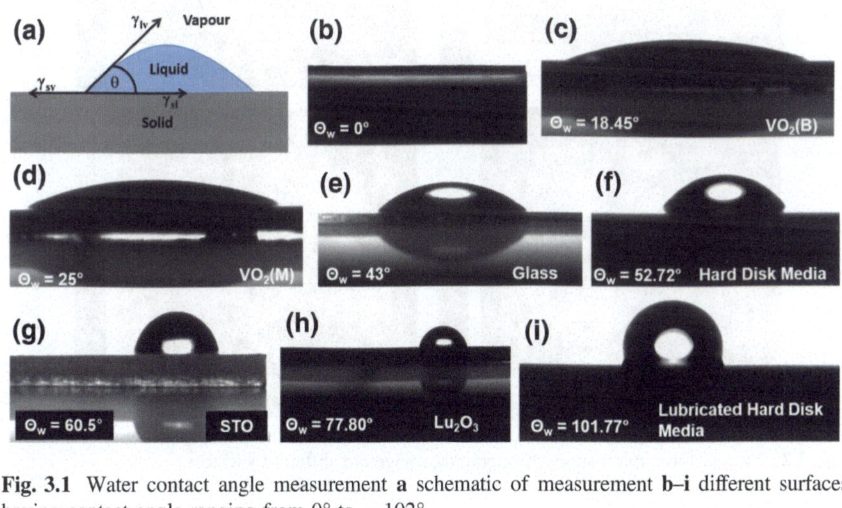

Fig. 3.1 Water contact angle measurement **a** schematic of measurement **b–i** different surfaces having contact angle ranging from 0° to ~102°

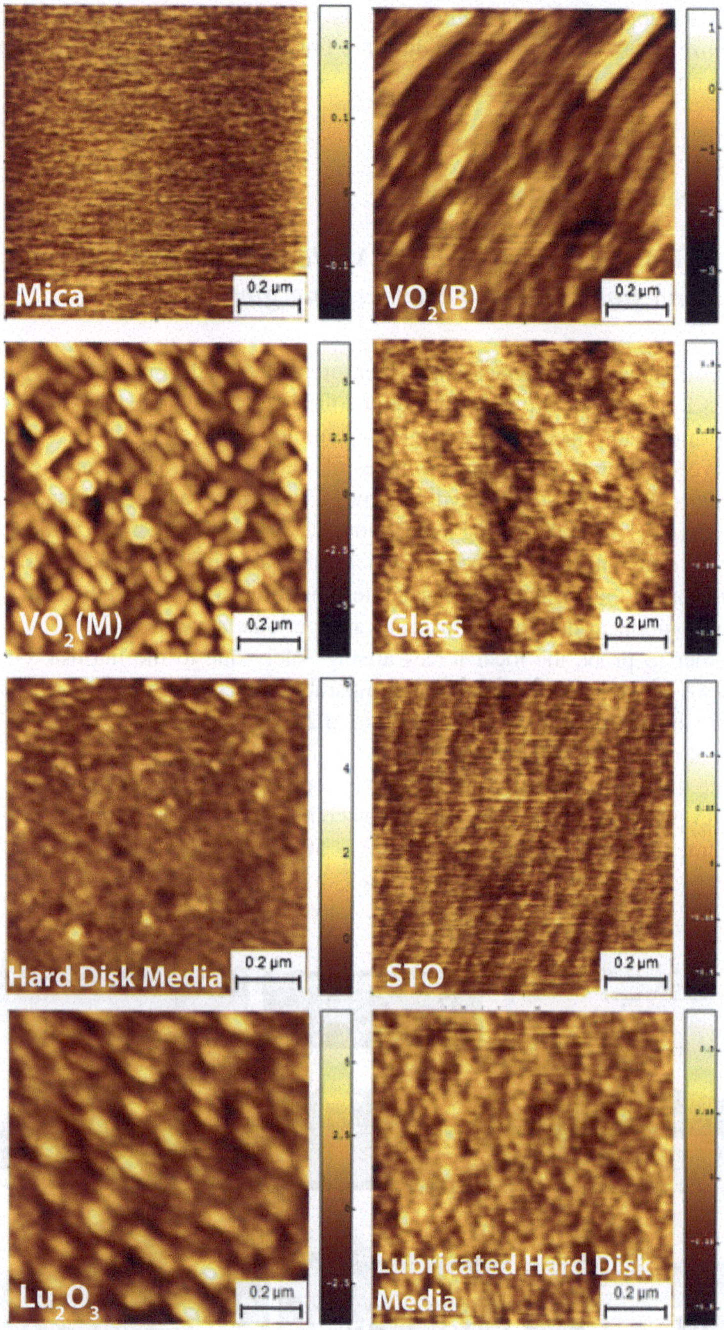

Fig. 3.2 Atomic force microscopy topographic images of different surfaces

Table 3.1 Contact angles on various surfaces obtained with three different liquids and calculations of surface energies with its two components

	Contact angle (degrees)			Surface energy (T)	Surface energy (D)	Surface energy (P)	Convergence
	Water	Ethylene glycol	Diiodomethane				
Mica	0	20.4	45.7	65.9	15.8	50.2	0.93
VO_2 (B)	18.4	16.0	27.3	61.6	24.0	37.6	0.90
VO_2 (M)	25.0	5.5	20.5	60.3	27.4	32.9	0.91
Glass	43.0	27.3	41.4	51.0	23.1	27.9	0.94
Hard disk media (CoCrPt magnetic layer)	52.7	49.3	44.0	42.8	22.4	20.4	0.86
$SrTiO_3$ (STO)	60.5	15.7	31.7	47.4	34.8	12.6	0.99
Lu_2O_3	77.8	63.1	49.5	33.1	28.1	5.0	0.84
Lubricated hard disk media	101.8	89.0	87.7	14.0	11.1	2.9	0.97

introduce Cassie-Baxter heterogeneity [8]. The observed roughness variations in our experiments (0.2–3 nm) are far too small to influence the contact angle. To further substantiate our point, we would raise the case of VO_2 (M phase) and VO_2 (B phase). The film of VO_2 (M phase) has a higher RMS roughness ($\sim$2 nm) than VO_2 (B phase) (RMS roughness $\sim$0.6 nm) but their contact angles are $\sim$25° and 18.5° respectively. If roughness was influencing the contact angle, VO_2 (M phase) should have had a lower contact angle as compared to VO_2 (B phase). With roughness out of the way, it is now appropriate to claim that it is the surface chemistry and atomic structures which are contributing to the macroscale contact angle and surface energy in this study.

As per Fowkes method, these interactions can be interpreted as the geometric mean of a dispersive and polar component of surface tensions σ^D and σ^P respectively. The solid/liquid interfacial energy can be written as

$$\gamma_{sl} \approx \sigma^s + \sigma^l - 2\left(\sqrt{\sigma_s^D \cdot \sigma_l^D} + \sqrt{\sigma_s^P \cdot \sigma_l^P}\right)$$

Therefore, two or more liquids with known polar and dispersive components are used to find the solid surface free energy using OWRK model. The OWRK method separates the interfacial surface tension according to the underlying interactions (polar and dispersive) between the molecules. Hence, the total surface energy of the solid is the sum of polar and dispersive components. The polar component originates from the permanent dipole-dipole interactions. They are stronger in polar molecules (having permanent dipole moment). On the other hand, the dispersive component arises due to random fluctuations in the electron density when brought together, leading to an induced dipole-dipole interaction (London forces). The detailed description of OWRK model and fitting results are shown in supplementary. Importantly, the convergence factor (indicated in Table 3.1) is very close to

~ 1, showing this model is very suitable to calculate surface energies with a high degree of accuracy.

3.2.2 Force Spectroscopy on Surfaces with Native Water Layer

Figure 3.3 shows the force spectroscopy (FS) measurements on these samples in ambient conditions. Figure 3.3a shows a schematic of the measurement. Very thin water layer [few atomic unit cells] is condensed between the AFM tip and the

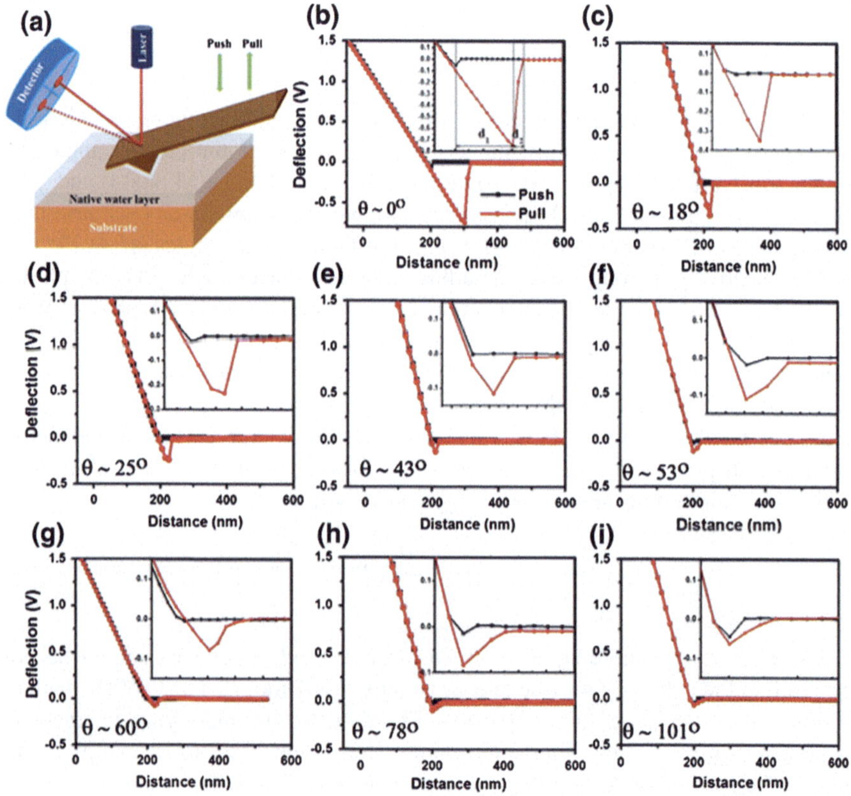

Fig. 3.3 Force spectroscopy (FS) measurements in ambient conditions, (**a**) schematic of measurement (**b–i**). FS curve on different surfaces having contact angle ranging from 0° to ~ 102°, (*b* mica, c VO_2(B) grown on STO, *d* VO_2(M) grown on STO, *e* cover slip glass, *f* hard disk media, *g* STO substrate, *h* Lu_2O_3 grown on YSZ, *i* lubricated hard disk media). *Inset* the curves plotted at a scale to clearly show the behaviour near detachment point. Distances before detachment from the surface (d_1) and the distance travelled by the probe in contact with the liquid meniscus (d_2) are shown in (**b**)

hydrophilic surfaces in ambient conditions [10, 16, 17]. The adsorbed water does not evaporate in capillary conditions between tip and sample at ambient conditions as per Kelvin's equation [22]. Figure 3.3b is a typical curve obtained on hydrophilic surface. The probe is first brought from non-interaction regime (showing no deflection of cantilever on detector) and pushed towards the sample (bending of cantilever due to repulsive forces showing positive deflection), and then retracted back to original position.

A detailed schematic illustration of the events that occur during the retraction of the tip has been given in Fig. 3.4. The negative deflection represents the attractive forces that hold the tip to the surface during retraction. The sharp detachment occurs when the pulling force of cantilever exceeds the intrinsic attractive force (Fig. 3.4b). Figure 3.2b–i shows the various force-distance curves obtained on a variety of samples with contact angle ranging from 0° to 102°. These figures show that the maximum pulling force reduces significantly as the contact angle increases. The maximum pulling force represents the tip-sample surfaces adhesive force occurring in Fig. 3.4b just before the tip breaks free from the liquid meniscus.

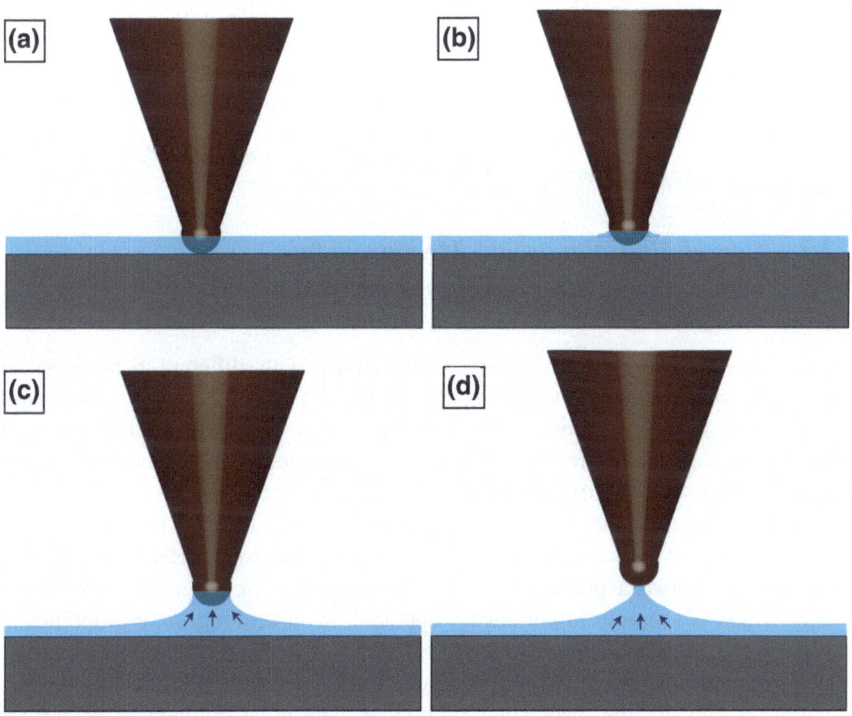

Fig. 3.4 Illustration depicting the various phases of tip interaction with the surface in the case of retraction (going away) from the sample. **a** Tip leaves the solid surface, **b** tip travels through the film of adsorbed water, **c** tip extends the neck of the meniscus formed by the water, **d** ultimately, the tip breaks free from the film of water

Indeed, there have been numerous studies on the adhesive forces over a period of time [11]. The adhesion force has been reported to be a function of humidity [18, 19] and the hydrophilic nature of surfaces [16]. Under ambient conditions, a meniscus is formed between the tip and a hydrophilic surface. The resulting capillary forces have been used to account for the pull-off forces in prior literature. Capillary forces are found to be stronger in hydrophilic surfaces [16–19]. However, numerical simulations along with some experimental results on mica surfaces have shown that the pulling force was incommensurately high compared to the expected capillary forces [16, 20, 21, 23]. Some classical theories do exist for tip-sample interactions: Hertz method assumes a linearly elastic sphere indenting an elastic surface but neglects the surface force and adhesion; JKR method is generally suitable for samples with high adhesion but ignores the long-range interactions outside the contact area and DMT method takes in to account the forces outside the contact area and is usually suited for samples with low adhesion. These theories have been developed considering simple van der Waals interaction. Particularly, the popular JKR model relates surface energy $\Delta\gamma \approx -\frac{2}{3\pi}\frac{F_{max}}{R}$, [11] where $\Delta\gamma$, F_{max}, and R, are surface energy, maximum pulling force and radius of tip respectively. However, capillary forces and pure van der Waal interactions have consistently failed to explain the inordinately high adhesion force observed in experimental results [20, 21]. Consequently, there is a gap in the understanding of the nature of adhesive and capillary forces intrinsic to hydrophilic surfaces. Indeed, there is insufficient experimental data on the correlation between adhesion forces and surface energy to validate the behavior of these models.

3.2.3 Correlating Nanoscale Measurements with Macroscale Measurements

Figure 3.5a–d shows the maximum pulling force with different parameters of macroscale contact angle measurements. Figure 3.5a shows that the maximum pulling force decreases exponentially with the contact angle. Most interestingly, it shows an exponential increase with the polar component of surface free energy Fig. 3.5b but not the dispersive component Fig. 3.5c.

As per JKR model, the maximum pulling force should increase linearly with the surface energy. We, however, observe an exponential trend. Here, we show a universal plot between pulling force versus the water contact angle and surface energy which is not only important for applications in probing these properties at nanoscale but also predicting the universal trend of adhesion forces with surface energy. Importantly, it is the polar component of surface energy that follows this relationship and not the dispersive component. The polar component of surface

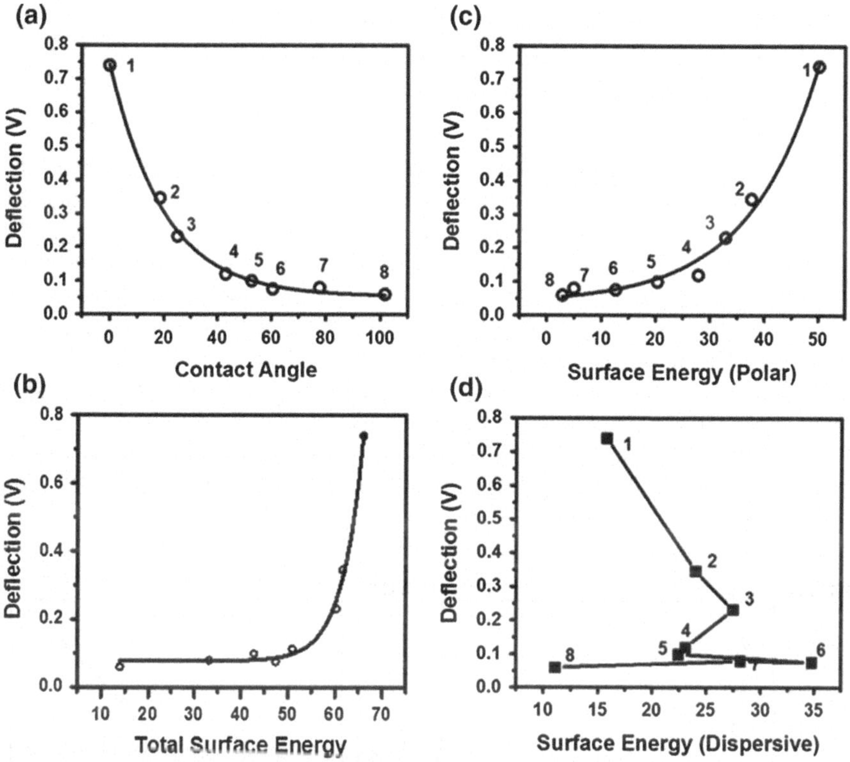

Fig. 3.5 The maximum pulling force plotted versus different parameters obtained from macroscale contact angle measurements, the maximum pull-off force versus **a** water contact angle **b** total surface energy **c** polar component of surface energy **d** dispersive component of surface energy (*1* mica, *2* VO_2(B) grown on STO, *3* VO_2(M) grown on STO, *4* cover slip glass, *5* hard disk media, *6* STO substrate, *7* Lu_2O_3 grown on YSZ, *8* lubricated hard disk media)

energy arises from the hydrogen bonding (which is due to electrostatic-dipole-dipole interaction) with the surface. In this case, the participating molecules and surfaces have a permanent dipole moment. On the other hand, the dispersive component of surface energy arises from "London forces" by inducing the instantaneous dipole moment, thereby giving a weak dipole-dipole interaction (van der Waals interaction). The scaling of the pull-off force with the polar component of surface energy and not the dispersive component provides clear evidence of electrostatic-dipole-dipole interaction being the key mechanism behind adhesive forces. Interestingly, in Fig. 3.3 the detachment is not sharp enough after the contact angle $\sim 25°$ and the total distance travelled in detachment process changes with contact angle of each surface.

3.2.4 *Breaking of Hydrogen Bonds on the Surface as a Possible Reason for Nanoscale-Macroscale Hydrophilicity Correlation*

Figure 3.6a, b are the distances travelled up to a point before (d_1) and after (d_2) the maximum negative deflection is achieved (refer Fig. 3.3b for graphical evaluation of d_1 and d_2), which essentially represents the pulling distance prior to the probe just leaving the surface and its travel in contact with liquid. The distance traveled by the probe (d_1) before detachment from the surface exponentially decreases with respect to water contact angle of the surface which is quite evident from the graph (Fig. 3.6a). On the other hand, the distance travelled in contact with the liquid meniscus increases linearly Fig. 3.6b. Hydrogen bonding forces have been found to decay exponentially compared to van der Waals forces that follow the power law [24]. However, it is difficult to distinguish between the two possible scenarios around the breaking of the hydrogen bond—it may break either from the solid

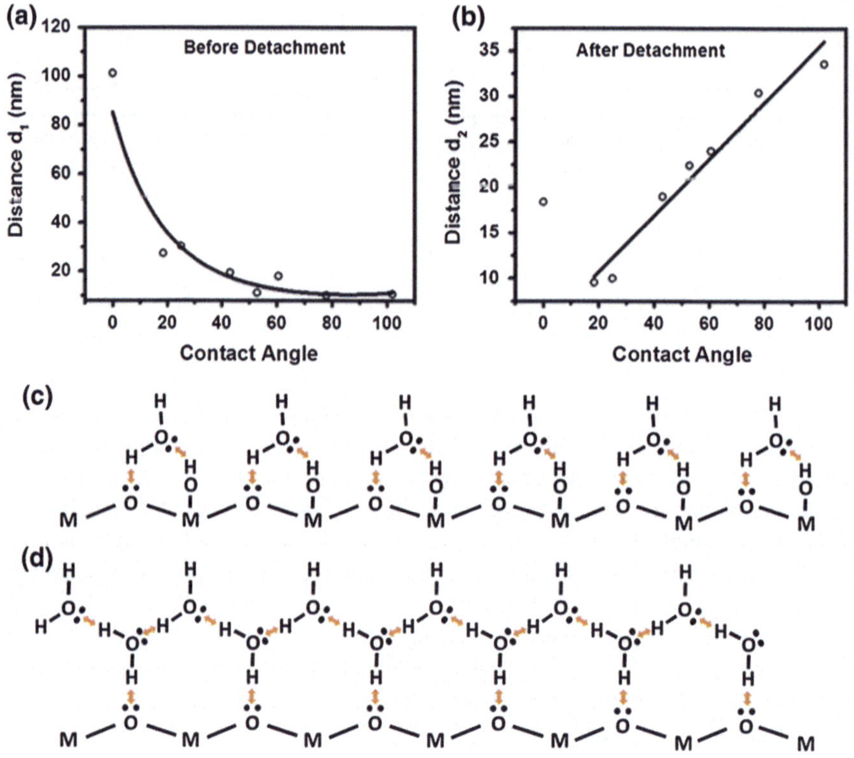

Fig. 3.6 **a**, **b** Are the distances travelled up to a point before and after the maximum negative deflection is achieved respectively. **c**, **d** Schematic showing hydrogen molecule structure at the interface of highly hydrophilic surface and relatively less hydrophilic surface respectively

surface or from a few molecular layers above the surface. At least, in the case of mica, we think it is more likely to have the breakage of the hydrogen bond occur a few molecular layers above the solid surface. Both the surfaces in question (the AFM tip and mica) are highly hydrophilic and hence, a rather large meniscus will be formed. We would like to point out that the molecular orientation of water layer at the interface can play an important role in breaking the number of hydrogen bonds using the AFM tip in both cases (breaking from solid surface or few atomic layer above it). The number of hydrogen bonds at the surface is decided by surface atomic structure and chemical nature of different hydrophilic and hydrophobic surfaces.

In order to elucidate the nature of water interaction with surfaces of varying degrees of hydrophilicity, we performed density functional theory (DFT) calculations on selected surfaces.

(Disclaimer: The simulation studies weren't performed by the candidate but are shown for completion. They were performed by Kelsey Stoerzinger, Yueh-Lin Lee, Livia Giordano and Yang Shao-Horn, Massachusetts Institute of Technology as part of our collaboration. The data has not been shown here. Results have been briefly summarized).

On the clean surfaces of transition metal oxides, VO_2(M)(022) and $SrTiO_3$(001), it is favorable for water to dissociate and hydroxylate the surface. On the more ionic Lu_2O_3 water adsorbs molecularly intact at fully coordinated metal ions, while it dissociates in proximity of under-coordinated Lu^{3+} ions also present at the surface of unreconstructed (111) surface of cubic Lu_2O_3. On the VO_2 surface, the hydroxyl groups are found to be stable, while a less favorable adsorption is found on $SrTiO_3$(001). The measured contact angle scales inversely with the adsorption free energy of hydroxyl groups The presence of OH groups on the surface, expected to scale with adsorption energy, introduces polar groups on the surface which can bind adsorbed water via electrostatic-dipole-dipole interactions [15] (Fig. 3.6c). Thus, the correlation of nanoscopic forces with the macroscale contact angle and surface energy (Fig. 3.5) can be explained by considering breakage of the number of hydrogen bonds specific to a molecular structure on each surface due to its inherent atomic structure and chemical composition. The detailed analysis of molecular structure can also explain Fig. 3.6a, b. In the case of more hydrophilic surfaces, a greater number of hydrogen bonds are formed perpendicular to the surface (Fig. 3.6c). On the other hand, there is a significant number of in-plane hydrogen bonds in the case of less hydrophilic surfaces (Fig. 3.6d). Therefore, when the probe is removed perpendicular to the surface a sharp detachment occurs when the pulling force exceeds the hydrogen bonding. This characteristic interaction force should exponentially decrease with contact angle as is seen in Fig. 3.6a. In Fig. 3.6b, we plot the distance travelled after detachment from the surface against the contact angle observed. On close observation, we find that the superhydrophilic surface (mica) differs from the trend. This is because in the all other cases, the tip being more hydrophilic than the surfaces being probed, encourages water to climb along the tip when the probe is retracted from the surface. Expectedly, the after detachment distance increases as the contact angle increases in Fig. 3.6b.

3.3 Conclusion

We have demonstrated that the adhesion forces are an exponential function of macroscale water contact angle and surface energy due to strong liquid-solid interaction involved in capillary forces. These forces are dominated by electrostatic-dipole-dipole interaction (hydrogen bonding) on the surface and/or few atomic water layers above the surface but not by London forces (induced-dipole-dipole interactions). This work clearly shows that molecular simulations of wetting using a small number of atoms typical of the atomic force measurements, may enable us to predict macroscopic wetting properties of surfaces and opens new avenues for fundamental understanding of interfacial phenomenon in technologies such as dip-pen lithography and probing wetting properties on the nanoscale. For example, we can use such an approach to locally evaluate surface energies and probe contact angles for nano patterned surfaces. In such a scenario, using the conventional water droplet method would result in obtaining a value which is an average of all local values.

Hence in this chapter, we have looked at how we can use the traditional force-distance curves to explain how water, the simplest biomolecule, interacts with unreactive inorganic surfaces, in the process enabling wetting measurements to be made in the nanoscale. In the next chapter, we will move to studying a more complex class of biomolecules lipids, typically weighing around hundreds of daltons, interacting with inorganic metal oxide surfaces.

3.4 Experimental Details

3.4.1 Sample Fabrication

1. **Sample 1** (mica) and **Sample 4** (cover slip glass) were commercially procured.
2. **Sample 2 and 3**: Monoclinic $VO_2(B)$ and $VO_2(M)$ thin films on (100) $SrTiO_3$(STO) substrates were grown using PLD respectively. The target was laser ablated using a pulsed KrF Excimer laser having a wavelength of $\lambda = 248$ nm and pulse width of approximately 12 ns. The laser energy density was fixed at ~ 2 J/cm^2 during the optimization of film growth parameters. The substrate temperature was kept 500 °C throughout the deposition. At low pressure (1×10^{-4} Torr) and 5 Hz laser rep rate, M phase is stabilized. At and above a pressure of 5×10^{-3} Torr and the laser rep rate to 2 Hz gave rise to a single phase $VO_2(B)$ film.
3. **Sample 5**: Commercial hard disk media is Ar+ ion etched to remove the conventional carbon overcoat. Hence the top surface is the CoCrPt based magnetic layer (Though the top carbon layer is removed, it may contain traces of carbon).

4. **Sample 6**: STO substrate was procured commercially from Crystec GMBH. The substrates were sonicated in DI water for 15 min. After that the substrates were kept in buffered HF solution for 30 s. They were subsequently washed with DI water and blow dried. The treated substrates were then annealed in the furnace at 950 °C for 2 h. The ramp up rate was 5°/min and the ramp down rate was 3°/min.
5. **Sample 7**: Lu_2O_3 film was grown using PLD technique. The target was laser ablated using a pulsed KrF excimer laser having a wavelength of $\lambda = 248$ nm and pulse width of approximately 12 ns. The laser energy density was fixed ~2 J/cm^2 during the optimization of film growth parameters. The substrate temperature was kept 750 °C throughout the deposition. The pressure was 10^{-6} torr throughout the deposition.
6. **Sample 8**: Commercial hard disk media is Ar+ ion etched to remove the conventional carbon overcoat and treated with high energy carbon using FCVA (Filtered Cathodic Vacuum Arc). Then the surface was treated with PFPE (Per fluoro polyether) based lubricant (Zdol 4000).

3.4.2 Contact Angle Measurement Setup

The macroscale contact angle and surface energy measurements were carried out using the video based fully automated Data Physics optical contact angle microlitre dosing system (OCA 40 Micro). Solvent drops (1 μl/drop) of DI water (more polar), ethylene glycol (intermediate polar and dispersive) and diiodomethane (more dispersive) with known surface tensions (refer Table 3.1) were dispensed using a Teflon coated motor driven syringe. The contact angles were measured at ambient conditions and a video was recorded (72 frames/s) for every dispensed solvent droplet. Any dynamic changes to the droplet on the surface can be precisely observed through this method. The contact angles with three different test liquids and surface free energies (polar and dispersive) were evaluated using the sessile drop technique and Owens, Wendt, Rabel and Kaelble (OWRK) model.

3.4.3 Atomic Force Microscopy and Force Spectroscopy

The atomic force microscopy and spectroscopy measurements were done in contact mode under ambient conditions using a commercial system by Agilent (Model 5500) and the commercial tips by NANOSENSORS (Type PPP-CONT-10) of force constant ~0.02–0.77 N/m. The results are presented after repeating several measurements using many similar tips.

References

1. He, Y., et al.: Local ordering and electronic signatures of submonolayer water on anatase TiO_2(101). Nat. Mater. **8**(7), 585–589 (2009)
2. Azimi, G., et al.: Hydrophobicity of rare-earth oxide ceramics. Nat. Mater. **12**(4), 315–320 (2013)
3. Kuna, J.J., et al.: The effect of nanometre-scale structure on interfacial energy. Nat. Mater. **8**(10), 837–842 (2009)
4. Ostuni, E., et al.: Adsorption of proteins to hydrophobic sites on mixed self-assembled monolayers†. Langmuir **19**(5), 1861–1872 (2003)
5. Vezenov, D.V., et al.: Chemical force spectroscopy in heterogeneous systems: intermolecular interactions involving epoxy polymer, mixed monolayers, and polar solvents. J. Am. Chem. Soc. **124**(35), 10578–10588 (2002)
6. Yao, X., Song, Y., Jiang, L.: Applications of bio-inspired special wettable surfaces. Adv. Mater. **23**(6), 719–734 (2011)
7. Nishimoto, S., Bhushan, B.: Bioinspired self-cleaning surfaces with superhydrophobicity, superoleophobicity, and superhydrophilicity. RSC Adv. **3**(3), 671–690 (2013)
8. Sun, T., et al.: Bioinspired surfaces with special wettability. Acc. Chem. Res. **38**(8), 644–652 (2005)
9. Piner, R.D., et al.: "Dip-pen" nanolithography. Science **283**(5402), 661–663 (1999)
10. Hong, S., Mirkin, C.A.: A nanoplotter with both parallel and serial writing capabilities. Science **288**(5472), 1808–1811 (2000)
11. Mittal, K.L.: Advances in Contact Angle, Wettability and Adhesion. Wiley (2013)
12. Das, S.C., et al.: Determination of the polar and total surface energy distributions of particulates by inverse gas chromatography. Langmuir **27**(2), 521–523 (2011)
13. Mante, P.-A., et al.: Probing hydrophilic interface of solid/liquid-water by nanoultrasonics. Sci. Rep. **4** (2014)
14. Giovambattista, N., Debenedetti, P.G., Rossky, P.J.: Enhanced surface hydrophobicity by coupling of surface polarity and topography. Proc. Natl. Acad. Sci. **106**(36), 15181–15185 (2009)
15. Giovambattista, N., Debenedetti, P.G., Rossky, P.J.: Effect of surface polarity on water contact angle and interfacial hydration structure. J. Phys. Chem. B **111**(32), 9581–9587 (2007)
16. Eastman, T., Zhu, D.-M.: Adhesion forces between surface-modified AFM tips and a mica surface. Langmuir **12**(11), 2859–2862 (1996)
17. Sedin, D.L., Rowlen, K.L.: Adhesion forces measured by atomic force microscopy in humid air. Anal. Chem. **72**(10), 2183–2189 (2000)
18. Xu, L., et al.: Wetting and capillary phenomena of water on mica. J. Phys. Chem. B **102**(3), 540–548 (1998)
19. Xiao, X., Qian, L.: Investigation of humidity-dependent capillary force. Langmuir **16**(21), 8153–8158 (2000)
20. Jang, J., Schatz, G.C., Ratner, M.A.: Capillary force on a nanoscale tip in dip-pen nanolithography. Phys. Rev. Lett. **90**(15), 156104 (2003)
21. Jang, J., Schatz, G.C., Ratner, M.A.: Capillary force in atomic force microscopy. J. Chem. Phys. **120**(3), 1157–1160 (2004)
22. Thomson, W.: 4. on the equilibrium of vapour at a curved surface of liquid. Proc. Roy. Soc. Edinb. **7**, 63–68 (1872)
23. Burnham, N.A., et al.: Probing the surface forces of monolayer films with an atomic-force microscope. Phys. Rev. Lett. **64**(16), 1931–1934 (1990)
24. Pashley, R., et al.: Attractive forces between uncharged hydrophobic surfaces: direct measurements in aqueous solution. Science **229**(4718), 1088–1089 (1985)

Chapter 4
Quantifying Interactions Between Lipids and Surfaces

Abstract Sample preparation techniques remain one of the most prominent bottlenecks in the field of lipidomics, where the extremely skewed distribution of lipids by their class prevents the discovery and characterization of new lipid species in the lipidome. In this chapter, we study the interactions of phospholipids with surfaces made of metal oxides. Knowledge of such interactions are key to the fabrication of efficient solid phase extraction resins which can resolve the conundrum in sample preparation strategies. We present comparisons between various metal oxides with respect to their phospholipid enrichment capacities in 3 formats, first by modifying MALDI plates, second by fabricating μcolumns packed with metal oxide powder and finally third by employing a batch sample processing technique incorporating metal oxide powder. Thereafter, owing to the scatter in acquisition of data in all the former approaches, we demonstrate an extremely sensitive technique based on surface plasmon resonance to quantify phospholipid adhesion to metal oxide surfaces with sub-nanogram level accuracy. We also show the effect of the headgroup in determining the affinity to metal oxide surfaces. Finally, we demonstrate a new technique based on thin planar TiO_2 columns deposited on MALDI target plates to perform on-target enrichment on phospholipid species, in a format easily integrated into existing commercial MALDI-MS systems.

4.1 Introduction

Having discussed the interactions of water with various surfaces, we now move to a heavier class of molecules, which are very important in the state of human health. We are now going to describe studies on interactions of lipids with inorganic oxide surfaces. Since the transition from studying water to lipids is seemingly abrupt, we will provide a very detailed introduction to the class of biomolecules known as lipids and why we intend to study their interactions with oxide surfaces quantitatively.

In the interest of completeness of discussion, however, we will have to start with a larger field of study called metabolomics of which, study of lipids (lipidomics) is a part.

A. Patra, *Quantifying Interactions of Biomolecules with Inorganic Surfaces*, Springer Theses, DOI 10.1007/978-3-319-30728-2_4

4.1.1 Metabolomics

The scientists, then "philosophers" of ancient Greece had the notion that changes in biological fluids are indicators of disease. In fact, in the Middle Ages "urine charts" were used to map colors, smells and tastes of urine to various diseases. Since these changes are now known to be metabolic in nature, it is fair to say that these scientists were, in fact, practicing the earliest forms of metabolomics. A medieval chart is shown in Fig. 4.1.

As is, probably, evident from the nomenclature, metabolomics is the system level study of metabolism. More specifically though, metabolomics seeks to determine and quantify, through exhaustive methods, the presence of thousands of small molecules of interest (metabolites) in cells, tissues and organisms at large. The whole field of study revolves around the governing idea—the metabolic state,

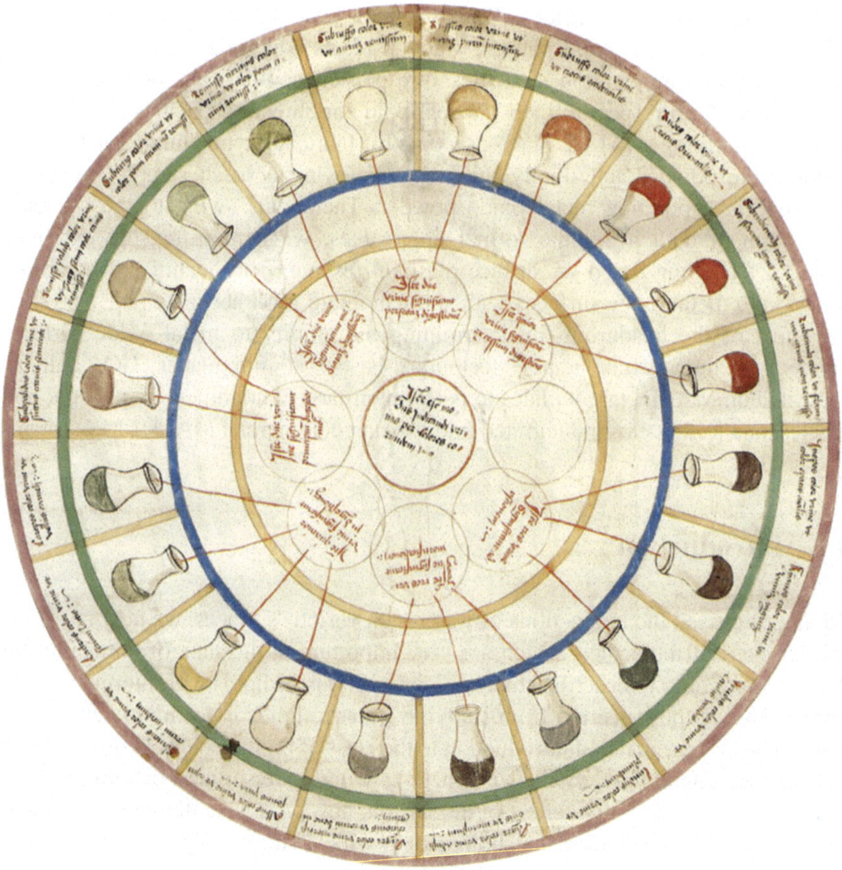

Fig. 4.1 A medieval urine chart detailing conditions that can be diagnosed from change in apparent properties of urine (*Courtesy* Scientific American)

at any given time, of a biological entity is the snapshot of the overall physiological state. Scientists in ancient China used ants to detect high levels of glucose in urine and thereby predicting diabetes. A thousand years later, metabolomic studies are able to capture global biochemical events by assaying thousands of small molecules in cells, tissues, organs, or biological fluids—followed by the application of informatics techniques to define metabolomic signatures. Metabolomic studies can lead to enhanced understanding of disease mechanisms and to new diagnostic markers as well as enhanced understanding of mechanisms for drug or xenobiotic effect and increased ability to predict individual variation in drug response phenotypes (pharmacometabolomics). The identities and concentrations of these metabolites result from a complex interplay among gene expression, protein expression, and the environmental interaction. Metabolomics makes use of instruments and technologies that can simultaneously quantify thousands of small molecules in a biological sample. The data obtained from these measurements must then be processed with sophisticated mathematical tools that can look for a molecular signature among millions of sets of data. Since, it is widely accepted that disease disrupts metabolism and consequentially causes changes that are long lasting, we have an opportunity to capture the diseased state as abnormal metabolic signatures. In literature, metabolomic signatures have already been reported for several diseases, including motor neuron disease [1], depression [2], schizophrenia [3–5], Alzheimer's disease [6], cardiovascular and coronary artery disease [7, 8], hypertension [9], subarachnoid hemorrhage [10], preeclampsia [11], type 2 diabetes [12–14], liver cancer [15], ovarian cancer [16], breast cancer [17], and Huntington's disease [18]. It is important to clarify the meaning of these "signatures" at this point. The concentrations of metabolites are modified in the state of disease and after treatment with drugs—hence the new signatures. Hence, with analysis, these signatures can provide information relating to disease pathophysiology. It isn't a stretch to say that metabolomics is going to have deep implications for both basic biomedical research and medical practice because it can capture vital information pertaining to active mechanisms of disease and also of drug action, thereby making it possible to correlate risk of disease and drug action to pathways of metabolism.

Metabolomics is an umbrella term because of the large number of metabolites in a typical organism. However, the sub-set which has captured the imagination of scientists in the last decade is lipidomics.

4.1.2 Lipidomics

The reaction of various fatty acids with backbone structures (such as, glycerol and sphingoid bases) give rise to a huge number of chemically distinct molecular species. It is believed that the lipidome of the eukaryotic cell contains more than 100,000 different lipid entities. Most remarkably, their function is as varied as their structure and chemical composition. Lipids regulate structure and chemical composition of cell membranes, store energy or become precursor to metabolites [19, 20].

Lipidomics is the systems-level study and analysis of lipids and factors that interact with lipids. It aims to analyze cellular pathways and networks by a comprehensive quantitative definition of lipidomes of biological entities—organelles to cells to organisms. In order to understand the interaction of lipids with biological factors in an organism, the first and foremost step is to extract the lipids from the tissue or cell. This complex lipid mixture is then "processed" to enable its analysis by various analytical techniques. This leads to the derivation of the 'lipid profile'. This so called lipid profile contains information pertaining to the lipid composition and abundance of lipid species in the extract. As is common in the field of proteomics and genomics, one tries to understand which lipids show variation in their levels of presence (upregulation/downregulation) as and when we perturb the biological homeostasis of the system under investigation.

For a very long time, lipids have been viewed as building blocks of the cell membrane with absolutely no other function other than structural. It is a fact that the ultimate macromolecular architecture of lipid assemblies, are decided, at least in part, by the geometrical properties of individual lipid molecules. For clarity, it is prudent to discuss the case of phosphatidylcholine and phosphatidylethanolamine. The size and orientation of the choline headgroup of phosphatidylcholine forces the fatty acid chains to contort into a geometric shape resembling a cylinder. Hence, phosphatidylcholine molecules, when lined up next to each other, will adopt a bilayer arrangement. Phosphatidylethanolamine, however, tends to form a hexagonal shape by itself due to its reduced headgroup size and state of hydration. The shape and packing of lipid molecules in macromolecular assemblies are important factors having significant effects on the dynamics of membranes—for example, in intermediates of membrane fusion and in the formation of specialized microdomains [21].

Precise determination of molecular lipid species becomes important not only in lieu of understanding their individual roles and functions but also to shed light on their distinct behavior in a lipid collective. Ultimately this will open up new avenues in cell biology, biochemistry and biophysics. The fact that lipidomics is going to answer questions relating to the complex lipid metabolism machinery and its relation to the construction of new cellular membranes is also going to be of paramount importance to our understanding of human diseases. It is already known that atherosclerosis, infectious diseases, Alzheimer's disease and cancer all have a lipid component in their epidemiology. As the mechanisms of action are understood, it will unravel the main culprits in the course of a disease affliction. This will lead to better targeted drugs aimed at the cause, rather than the symptoms.

Lipids are also biomarkers, which means that even before we start discussing treatment of diseases, it will be interesting to point out the potential of lipids to be diagnosis tools. Lipid levels are tightly regulated, spatially and temporally, in the various parts of the human body. It is not outlandish to expect that deregulation of this network contributes to the onset of a pathological condition. This idea is indeed supported by a wide variety of human diseases (as pointed out above) that are associated with abnormal lipid metabolism.

In the field of nutrition research, an understanding of lipids is going to enable innovation of healthier food formulas. Furthermore, it will be possible to engineer nutrition specific to people with particular needs.

The applications are practically endless.

This brings us to the motivation of this study. One of the technological bottlenecks in lipidomics revolves around sample preparation. Unfortunately, the improvements in instrumentation which have produced increasingly sensitive mass spectrometers has not been complemented by similar advances in separation science. In essence, what we are trying to do, thereby, is to better sample preparation for mass spectrometry.

4.1.3 The Need for Sample Preparation

There are reports in literature about metabolomic profiling dating as far back as 1960–70. But technological advances in instrumentation combined with the need for multiple levels of profiling (in addition to proteomics and genomics) has renewed interest in the field. While far more sophisticated instrumentation has pushed the limits of detection by orders of magnitude, sample preparation and analytical procedures at large have remained the technological bottleneck in the domain. Consequently, it is an active area of research and a subject of many reviews in literature. The choice of sample preparation strategy is absolutely crucial to the success of the experiment. Surprisingly, it has also been one of the overlooked sections of the field. Obviously, the choice of the sample preparation strategy will influence the observed metabolite profile and quality of data obtained. If the capabilities and limitations of the method is not understood clearly, there is a definite possibility that the accuracy of the biological interpretations or conclusions drawn from the data is compromised. Considering the chemical diversity of the metabolome, there can be no one perfect strategy for sample preparation which ensures no bias in the processed sample towards certain groups in the metabolome. In this thesis, we will focus only on sample preparation for lipidomics in particular.

The ultimate goal of the sample preparation step is to reproducibly transform the sample of interest into a "format" which is compatible with the LC-MS analysis approach while keeping the original composition unaltered as much as possible. So if we were to describe the desired qualities of a sample preparation strategy, they would be

1. Simple and fast
2. Minimum number of steps
3. Compatible with automation
4. Reproducible
5. Cheap.

There are two major types of strategies which are employed in the preparation step—Liquid-liquid extraction (LLE) and Solid Phase extraction (SPE). We will discuss LLE and its disadvantages which dictate the need for SPE. Thereafter we are going to describe SPE and its details.

4.1.3.1 Liquid-Liquid Extraction (LLE)

The working principle of LLE is that a sample is partitioned between two immiscible liquids (also called phases) in the ratio of its solubilities in the respective phases. Most commonly, one phase is aqueous and the other is organic. Hence the more polar hydrophilic compounds separate into the aqueous phase and the non-polar hydrophobic compounds separate into the organic phase.

The distribution coefficient K_{dist} describes the partition of a particular compound between two immiscible solvents. The process of separation of a sample X between an aqueous and an organic solvent can be written as

$$X(aq) \leftrightarrow X(org)$$

$$K_{dist} = \frac{[X_{org}]}{[X_{aq}]}$$

where X_{org} and X_{aq} are the activities of X in the two solvents.

Ideally, K_{dist} needs to be as high as possible so that high degree of separation can be attained in a single run. However, if K_{dist} is not high enough, we can use continuous LLE in which the sample is subject to multiple runs of the LLE as illustrated above with fresh organic solvent every time.

With a basic understanding of the process of LLE, we are now in a position to discuss its limitations which in turn are reasons to move to SPE.

1. **Limited selectivity**—LLE works on the basis of differences in solubilities of analytes and the unwanted contaminants in aqueous and organic solvents. Since most lipids are soluble in organic solvents, all lipids are co-eluted together in the organic phase. This means that the extract will have all non-polar interferences and will be less pure than we want it to be.
2. **Large amount of solvents required**—It is necessary to use an adequate amount of solvents to extract all the lipids of interest from the complex sample. It is desirable to use as less of these solvents as possible because more often than not, they are both expensive and not environment-friendly.
3. **Emulsions**—This occurs when the sample being purified has a large amount of surfactant like compounds (notably phospholipids). Having intermediate solubility in both phases, these compounds will form a milky mid-band between the two phases. This will prevent clean collection of either phase. Essentially, this renders LLE a less dependable method.

4. **Difficulty of automation**—Due to the large amount of solvents typically involved in successful runs of LLE, it is not easily compatible with automated sample handlers. This implies that LLE procedures will have problems of user bias and low throughput.

4.1.3.2 The Problem with Phospholipids

In the context of lipidomics, it is important to discuss the limited selectivity of the LLE procedure in detail as this is perhaps the cause of the bottleneck in the sample preparation sub-domain. Before describing the problem in detail, it is necessary to take a look at the distribution of lipids in human serum (Table 4.1).

Table 4.1 Lipid distribution in the plasma (from Quehenberger et al. [37])

Lipid category	No. of species	Sum (mg/100 ml)
Fatty acyls		
Fatty acids	31	5.82
Eicosanoids	76	0.002
Glycerolipids		
Triacylglycerols	18	90.6
1,2-Diacylglycerols	28	2.36
1,3-Diacylglycerols	27	0.805
Glycerophospholipids		
PE	38	32.7
PC	31	157
PS	20	0.559
PG	16	0.480
PA	15	0.173
PI	19	2.74
N-Acyl PS	2	0.001
Sphingolipids		
Sphingomyelins	101	22.817
Monohexosylceramides	56	0.180
Ceramides	41	0.732
Sphingoid bases	6	0.02029
Sterol lipids		
Free sterols	14	31.8
Esterified sterols	22	114
Prenol lipids		
Dolichol	6	0.003
Coenzyme Q	2	0.394

It is hence clear that a massive chunk of the lipidome is made up of various subclasses of phospholipids. This heavily skewed distribution leads to prejudice in low molecular weight (LMW) metabolite coverage. Hence biological variations in the levels of these LMW metabolites get masked out by the overwhelming presence of phospholipids. In the absence of proper mobile phase, these phospholipids accumulate in the column, thereby accelerating its degradation. In addition, these accumulated species are slowly released in subsequent runs with other metabolites which leads to gradual increase in baseline noise levels, induction of matrix effects and alteration of separation characteristics. Phospholipids are also known to cause ion enhancement and suppression, which implies that the mass accuracy of co eluting metabolites are severely affected.

It is clear that LLE as a procedure will be of limited help in processing samples in such a way that selective lipid fractions are enriched and can be analyzed separately. All lipid species which are more soluble in organic solvents will be collected together and as described above, will cause a lot of problems both in experimental runs and in result analysis.

Due to these limitations imposed by the fundamental working principles of LLE, we will look at SPE as an alternative sample preparation strategy.

4.1.3.3 Solid Phase Extraction

Like LLE, SPE is also a sample preparation scheme for chromatography. The most common configuration is to have a solid phase material (called solid phase extraction "resin") packed into a column between one or two frits. The liquid mixture which is desired to be purified is passed through this column. Due to chemical interactions between components in the mixture and SPE resin, some components will pass through and some will be retained. The bound components can be released or "eluted" by reversing the chemistry which led to their capture in the resin bed.

A pressure differential required to draw the liquid through the column can be generated by a small gas pressure on the top of the column or a vacuum at the base of it. Typically, a column would contain anywhere between 50 mg and 10 g of resin. Miniaturization is fairly straightforward with SPE. Due to this, SPE is particularly suited for situations where the volume of sample involved is very small.

4.1.3.4 Motivation

A number of materials have been explored for their possible application as an SPE. Hydrated silica [22], florisil [23], hydroxyapatite [24], hydrated metal oxides [25] (aluminum, magnesium, zirconium, titanium) as SPE resins for various domains have been reported in literature [26–28].

Sample preparation is a bottleneck and unresolved problem in lipidomics research. The choice of sample preparation method is crucial to the observed

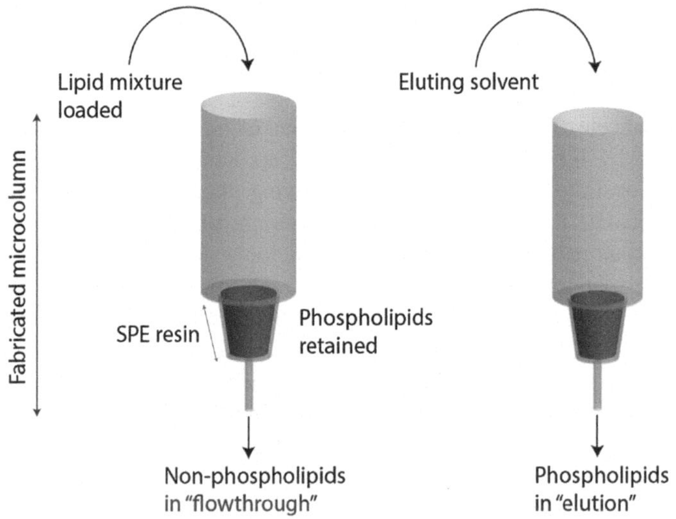

Fig. 4.2 A schematic illustration of the concept of SPE applied to lipidomics

lipidomics profile and the quality of data collected [29–31]. In the absence of a well-defined and understood sample preparation protocol, accuracy of biological correlative interpretation of collected data can be irreversibly compromised. We want to separate the phospholipids which are over-represented by mass in the lipidome (the collection of all lipid species in the lipid extract) from the rest of the lipidome. The process of electron spray ionization (described in Sect. 2.7.2) is a competitive phenomenon. The overwhelming presence of phospholipids in the lipid extract suppress the ionization of all other underrepresented lipid species. Hence, by removing the interfering class of lipids, phospholipids, it may be possible to discover more lipid species. The scheme is illustrated in Fig. 4.2.

When the mixture of lipids is passed through such a column it separates the components into fractions. In other words, the two fractions are now "enriched" because the interfering fraction has been removed. Hence, the name given to the process of separation is lipid fraction enrichment. Metal oxides are generally chemically resistant and abundant in the earth's crust which means there is a possibility of fabricating very cheap columns for lipid enrichment. We believe that there are 3 major criteria for a metal oxide to be chosen as an effective and commercially viable SPE resin.

1. **Stable and preferably non-toxic in ambient conditions**—some oxides like CoO and Cr_2O_3 are hygroscopic. On exposure to air, they will absorb moisture and make hydrated complexes. This renders them unusable in any column format. CrO_3 is carcinogenic making it a risky candidate to explore.
2. **High phospholipid capturing capacity**—The selected SPE resin must capture the highest amount of phospholipid per unit available surface area.

3. **Cheap**—If we can find such oxides which are abundant, there is a possibility of making extremely cheap, one time (or even multiple) use SPE cartridges packed with powders of said oxides.

The idea of selective phospholipid enrichment using TiO_2 was first proposed and demonstrated by Ikeguchi [32] and about 9 years later by Calvano [33]. Calvano actually demonstrated the possibility of removing phospholipids from dairy products and chocolate using TiO_2 columns. A more recent study [34] has shown the use of TiO_2 nanoparticles in analyzing the lipidome of olive oil and fruit.

To the best of our knowledge, there are no comparative studies in literature exploring the capturing capacities of different oxides under identical conditions so that chemistry of surface alone determines the capture amount. We seek to develop such protocols enabling accurate comparisons between various oxides so as to identify the best candidate for a cheap and specific SPE resin.

4.2 Results and Discussion

The development of protocols and methods for accurately measuring phospholipid to transition metal oxides went through a few iterations before we could develop sensitive methods which enabled true quantification of adhesion to chosen surfaces.

4.2.1 Metal Oxide Coated MALDI Plates for Selective Phospholipid Enrichment

MALDI coupled to mass spectrometer (MALDI-MS) is a high throughput process for quickly testing the metal oxides we want to employ as SPE resins. We deposit stable thin films of metal oxides on commercially obtained MALDI target plates (made of stainless steel). The details of the deposition procedure can be found in the experimental Sect. 4.4.1.

Previously, Calvano et al. [33] have shown that TiO_2 films can capture phospholipids from lipid extract obtained from dairy products. We synthesized an artificial mixture of lipids using 2 phospholipids (DPMC and SM) and 3 non-phospholipids (GluCer, Cer 534, Cer 574) and tested the various oxide thin films for their performance in selectively capturing phospholipids. For clarity, we have illustrated the workflow of the experiment in the Fig. 4.3. Experimental details have been provided in Sect. 4.4.2.

The data depicting the comparison between various oxides is summarized in Fig. 4.4. In an ideal scenario, the before wash data should show all 5 lipid species which would reduce to 2 (only phospholipids) after being washed with a 0.1% TFA acidic buffer. In our experimental runs, we noticed that STO performed the best, selectively retaining the phospholipids while letting the ceramides be washed away.

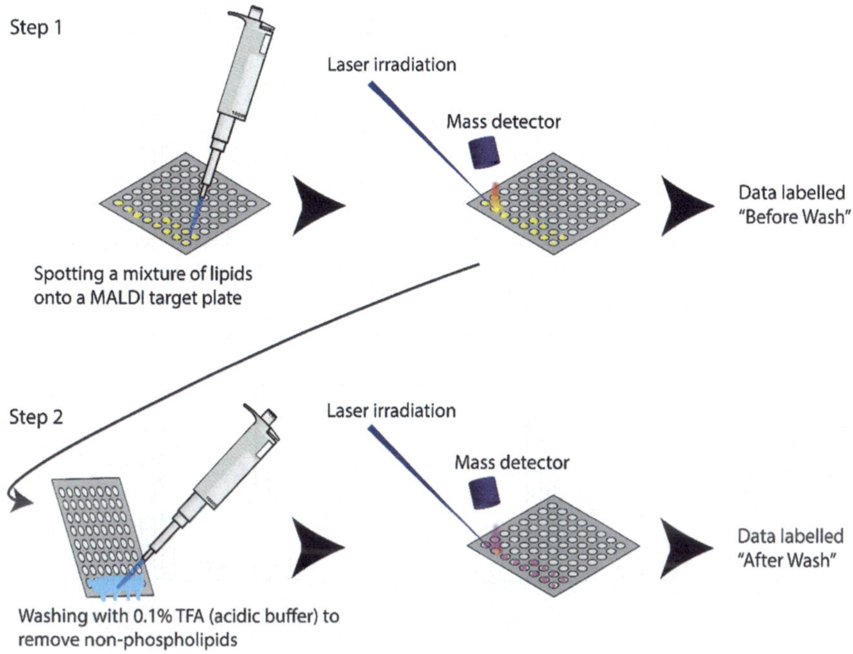

Fig. 4.3 Schematic workflow of the experiment for testing metal oxide thin films as SPE resins in the MALDI format

In some cases, such as LAO and Y_2O_3 the intensities of signal obtained before wash is actually smaller than the after wash signals. This is unexpected but can be explained by entering a short discussion about the intricacies of measuring lipid species through the MALDI-MS technique.

First of all, the nature of the surfaces is such that the mixture of lipids tends to spread instead of staying in the marked out circles (from where the MALDI setup will read the signal). Secondly, the same is true for the ClCCA +2,5 DHB matrix used to couple the energy of the laser to the compounds under investigation in order to ionize them. Consequently, the crystallization of the matrix is not uniform which in turn affects the readings obtained.

Due to the aforementioned reasons, the error between runs of the experiment was significant. It is therefore, important, to move the protocol in a direction which would make data more reproducible.

4.2.2 Phospholipid Enrichment Using μcolumns

MALDI is of course the method of choice when the most important parameter of the experiment in question is its ease and speed of operation. However, the more

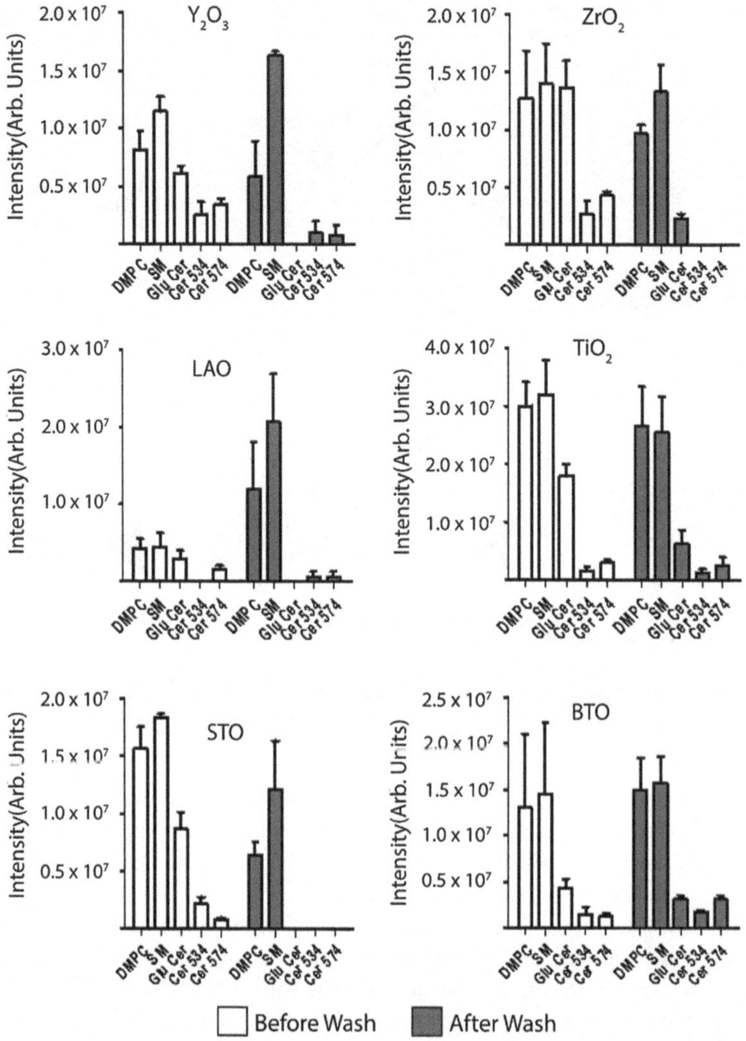

Fig. 4.4 Comparison between 6 chosen transition metal oxides with respect to their selectivity of enrichment of phospholipids DMPC and SM. Ideally, after a wash of MeCN and 0.1% TFA, the nonphospholipid fractions GluCer, Cer 534, Cer 574 should get removed from the plate leaving only DMPC and SM

realistic scenario in which these metal oxide powders will get deployed in, is columns. Hence the next iteration of seeking to study phospholipid binding to metal oxides involved fabrication of μcolumns. The process of fabrication is described in detail in Sect. 4.4.3. The experiment itself consists of a large number of steps, all of which have been set out in detail in the same Sect. 4.4.3.

There are a few issues that need attention when fabricating these μcolumns. We are using a suspension of metal oxide powders made in acetone to pack the μcolumn with these metal oxide powders. Since the amount of powder that can be packed within a small μcolumn is extremely small, there is probably significant differences in the actual weight of powders successfully packed into the μcolumn. Commonly available microbalances can't measure these weights reliably.

It is nearly impossible to fabricate working μcolumns with very fine powders. All the powders were commercially procured and they had large size dispersions. The SEM images of the powders are shown in Fig. 4.5. BTO, Sc_2O_3 and NiO powders had extremely small particles measuring few tens of nanometers. The particles pack so tightly together that it becomes impossible to push the liquid sample through them. The back pressure generated by the tight packing of these powders is so high that is most cases, applying a higher pressure to overcome it would lead to rupture of the column as such. Through repeated experiments we have found that for designing μcolumns with a packed bed of powders, the particle size has to be larger than 200 nm at the very least. The particle size of Y_2O_3 powders was about 200 nm and we managed to fabricate working μcolumns of the same with great difficulty.

We chose 4 oxides, STO, Y_2O_3, ZrO_2 and TiO_2 for testing relative efficiencies in selective capture of phospholipids from an artificial lipid mixture. The details of the lipid mixture can be looked up in Sect. 4.4.3. Briefly, we used a mixture of 2 phospholipids (DMPC and SM) and two non-phospholipids (GluCer and Cer 17). Ideally the flowthrough component should have only GluCer and Cer 17. The elution component must have only DMPC and SM. We use an acidic buffer to load the lipid samples into the μcolumns thereby promoting phospholipid adhesion. Thereafter we use a basic buffer to elute the bound phospholipids. The results are shown in Fig. 4.6.

As we can readily see, Y_2O_3 seemed to be the cleanest and most effective SPE resin which was extremely selective in capturing the phospholipids from the mixture and eluting them under the flow of basic buffer through the μcolumn. All other columns showed phospholipids in flowthrough component. This is known as "breakthrough" in the field of chromatography. "Breakthrough" happens when the maximum binding capacity of a column is exceeded.

It has to be pointed out that for the purposes of testing materials for SPE resins, we are using an artificial mixture of lipids in which phospholipids are present in a concentration about 1/10th of non-phospholipids. In actual lipid extracts, phospholipids are the dominant entity. The idea is to demonstrate that these SPE resins made from metal oxide powders can actually selectively capture phospholipids even when they are in a mixture where the major component is the non-phospholipid fraction.

Additionally, there are two factors which are not controlled in this setup

1. **Flow rate through the column**—This translates into the actual time the lipid species spend in contact with the metal oxide powders. Since manual pressure is used through a syringe and the powders vary a lot in their particle size, the flow rate of the liquid mixture through the packed bed of the column is different in each case.

Fig. 4.5 SEM images of powders showing the large difference in particle sizes. Scale bars are shown in *green*

2. **Surface area in contact with lipid species**—Owing to large differences in particle size and shapes of metal oxide powders, the actual surface area which is in contact with and can be used for enrichment of phospholipids is different.

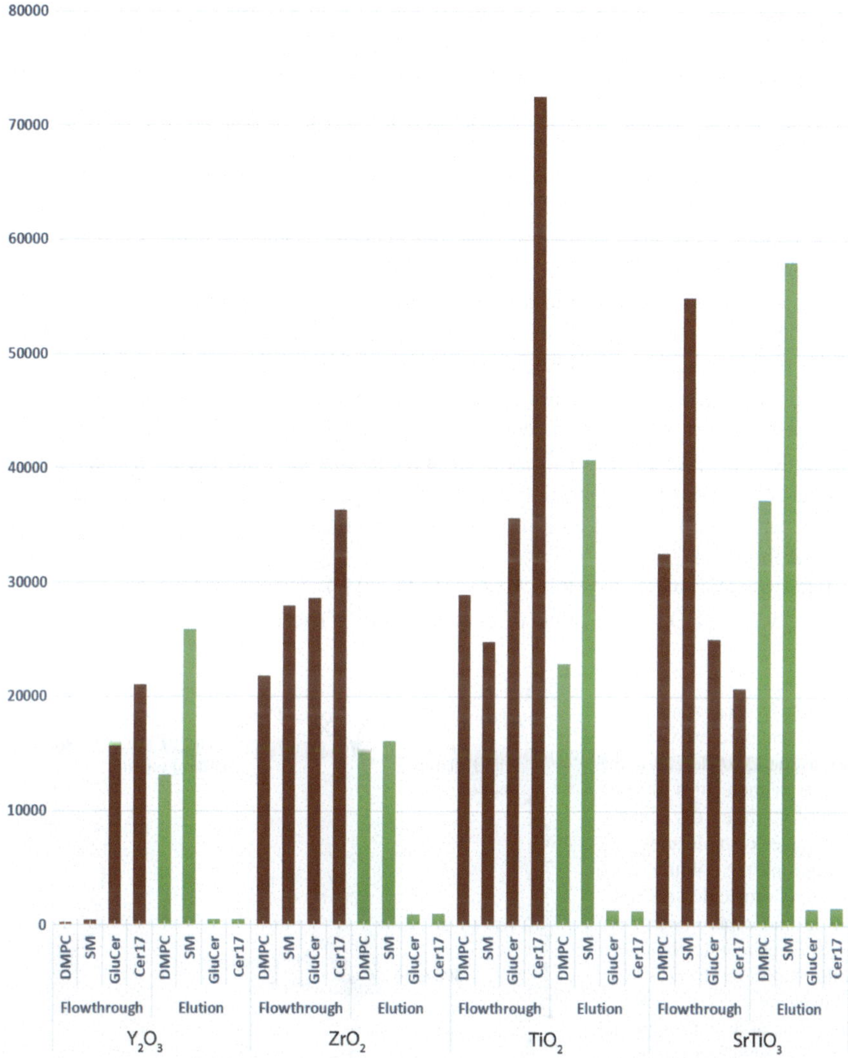

Fig. 4.6 Comparison between 4 oxides with respect to selective phospholipid enrichment. (*Brown* denotes flowthrough component, which should ideally have only GluCer and Cer 17; *green* denotes the elution component which should have only DMPC and SM. Y Axis is in arbitrary signal strength units

4.2.3 Phospholipid Enrichment Using Batch Processing

After identifying the major problem in fabricating and working with μcolumns packed with fine powders, we decided to test the various powders in a scheme which did not involve μcolumns. This approach is called batch processing. The details about the design of the experiment can be found in the Sect. 4.4.4. Through this approach, we have managed to ensure that the phospholipids spend the same amount of time in contact with the metal oxide powders. We also devised a parameter α which represents the ability of a particular metal oxide to act as a SPE resin. The details about the evaluation of α is schematically shown in Fig. 4.7.

The parameter α was evaluated for the phospholipid DMPC from a mixture of DMPC, SM, Glu Cer and Cer. When the experiment was performed for transition metal oxides from the 3d period, the broad trend could be studied. Figure 4.8 represents the variation of α in different oxides.

We can observe that α is around 0.7 for Sc_2O_3, V_2O_5, TiO_2, MnO and 0.8 for Fe_2O_3. NiO has an α of 0.4 while CuO and ZnO have 0.15 and 0.2 respectively. Hence it becomes immediately clear that for fabricating columns for phospholipid enrichment, usage of NiO, CuO or ZnO is futile.

Again, for the purposes of testing, we are using an artificial standard mixture where the non-phospholipid fraction is present in a ten-fold higher concentration compared to the phospholipid fraction. We want to be able to show that metal oxides can indeed work even in conditions biased heavily against selective capture of phospholipids.

Interestingly, we noted that Sc_2O_3, V_2O_5, TiO_2, MnO and Fe_2O_3 are all d^0 systems while NiO is d^8, CuO is d^9 and ZnO is d^{10}. Therefore, d^0 systems prefer the

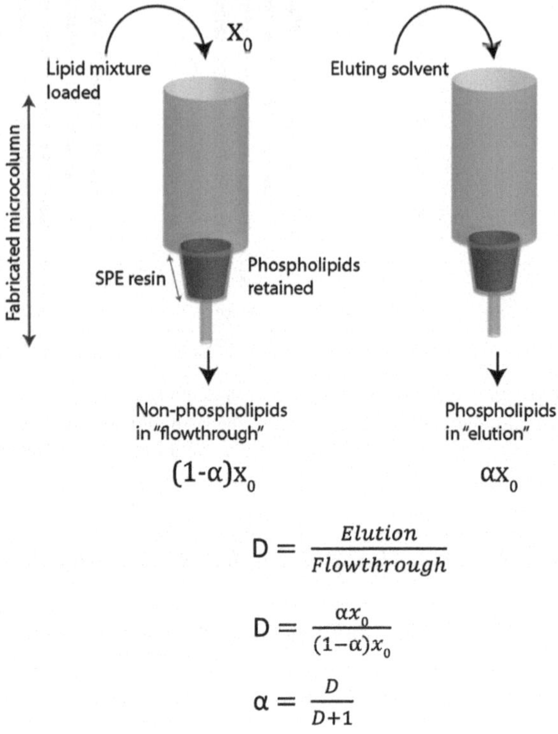

Fig. 4.7 Evaluation of α used to compare between various metal oxides. x_0 is the number of molecules in the original mixture. αx_0 denotes the number of molecules captured by the column. $(1 - \alpha)x_0$ molecules are not able to find capture sites on the SPE resin and are present in the flowthrough. D is obtained by dividing peak intensities of flowthrough and elution as obtained through ESI-MS. α will be 1 for a perfect SPE and 0 for a material which has no affinity for phospholipids

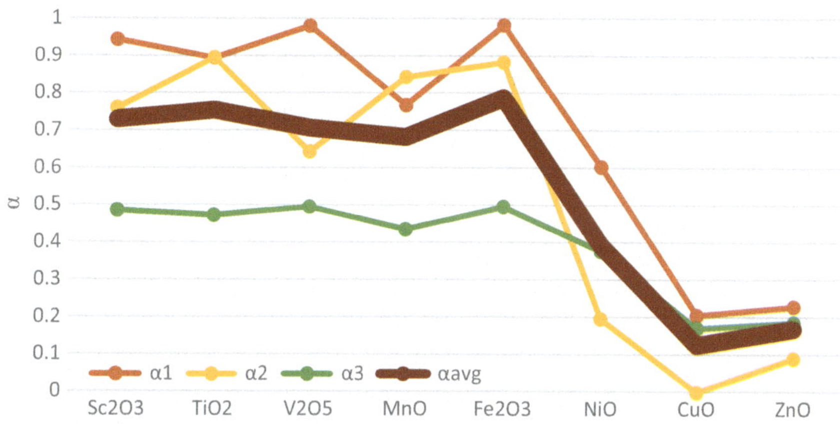

Fig. 4.8 α as obtained for all stable 3d transition metal oxides

adhesion of phospholipids and the adhesion is dramatically less for oxides where the d orbitals are filled with electrons. We can, hence, claim that with respect to choice of material for lipid fraction enrichment, one must try and choose metal oxides which have a large number of empty (if not completely empty) d orbitals.

Cr_2O_3 and CoO were not included in this study despite belonging to the 3d group because of their hygroscopic nature.

However, as seen from the plot, the spread in as-obtained α values about the mean was very significant. The standard deviation hovered around 30% of the mean value in most cases. Although moving to this scheme of testing SPE resins from the previous μcolumn based approach enabled us to control the time spent by lipid species in contact with the powder, there were still two crucial factors causing a sizeable spread across numerous runs of the same experiment.

1. **Large number of manual steps**—this introduces user dependent artefacts and increases the possibility of errors
2. **No control over surface area of particles**—again, as with the μcolumn approach, the actual surface area of the particles acting as SPE resin for phospholipids is different for various oxides and also amongst different runs of the same oxide.

4.2.4 Surface Plasmon Resonance for Quantifying Phospholipid Adhesion

Due to the lack of control over certain parameters as shown above, we decided to fabricate special chips which would enable us to measure accurately the amount of phospholipid molecules stuck to a thin film of metal oxide. The process of fabricating these chips has been described in detail in the Sect. 4.4.5. There are a number of issues which need to be carefully addressed in the fabrication of these chips. These have been discussed in detail in Sect. 2.5.4.

The objective behind moving to SPR setup for measuring phospholipid adhesion to metal oxide surfaces was the higher degree of control over experimental parameters and the possibility of minimizing errors inherent in manual steps.

1. Flow was controlled through a peristaltic pump having same fluid flow rate for every metal oxide chip
2. Surface area of SPE resin in contact with the phospholipid is the same for all samples

Now when these two parameters have been brought under our control, we are in a position to compare the various metal oxides through the surface plasmon resonance technique. Actual data obtained on a chip with a 3 nm film of TiO_2 is shown in Fig. 4.9 for the purpose of illustration of the concept. We can visualize every event in real time because of the nature of the experiment, unlike all other previous approaches where only the initial and final condition of the sample are known through readouts from the mass spectrometry step.

In Fig. 4.10, we show the shift of the resonance angle. The shift of the angle to the right on the binding of phospholipid and thereafter to the left on the unbinding is clearly seen. WINSPALL software package is used to fit the curves acquired experimentally. Using data from these simulated curves, the precise mass captured on the surface of the thin film is estimated. Data is shown for a film of TiO_2.

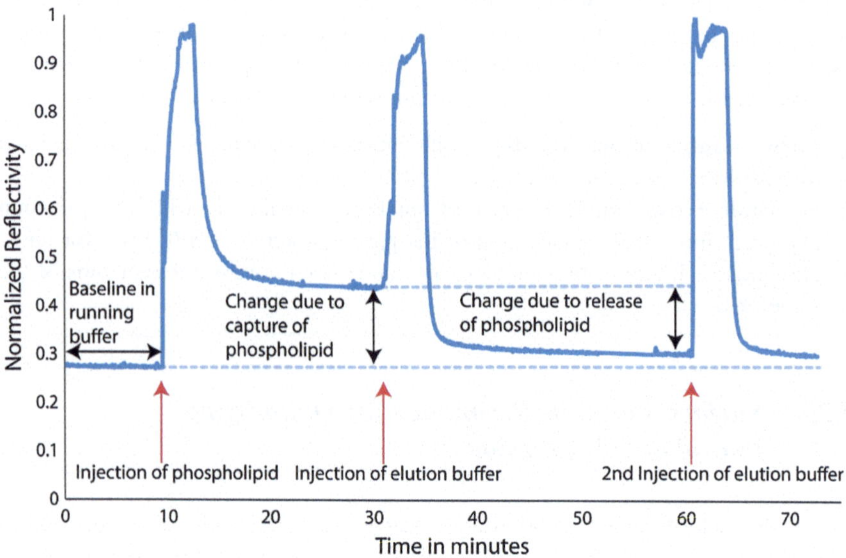

Fig. 4.9 The capture and release of the phospholipid DMPC from a 3 nm film of TiO_2 observed through surface plasmon resonance. Various events have been marked on the plot

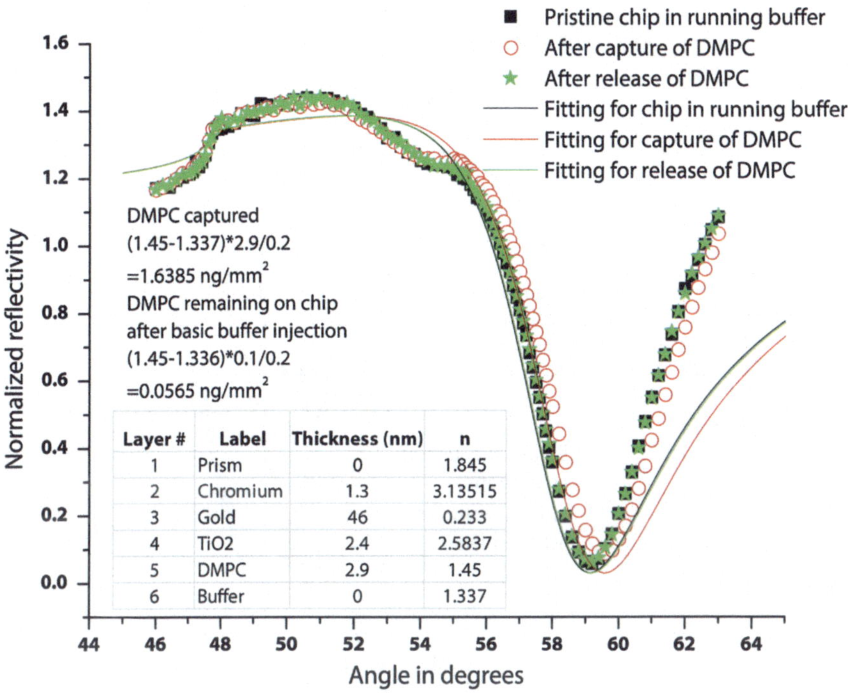

Fig. 4.10 Angular spectrum showing shifts in resonant dips upon binding and unbinding DMPC. The actual and fitted data is shown. The estimation of mass on the TiO_2 surface is derived and displayed in the inset. n is refractive index. Details of calculation can be found in Sect. 4.4.5

The extreme sensitivity of the SPR technique was found to be very useful not only in detecting phospholipid binding to the oxide surfaces but also in detecting etching away of the metal oxides themselves under the acidic buffer used to promote the capture of DMPC. When acids etch away at metal oxides, they form salts. The presence of salts are very detrimental to the system health of modern day sophisticated and precise mass spectrometry setups. The other serious problem is that salts cause ionization suppression [35, 36] and hence will lead to erroneous results. It is impossible to monitor the loss of metal oxides from μcolumns as salts because the change in mass is minuscule. However, with the SPR approach, the process can be easily detected. We show an example in Fig. 4.11 where the film of CuO is being etched by the 0.1% TFA acidic buffer used to inject DMPC.

We found that Sc_2O_3, V_2O_5, CuO and ZnO, having considerable basic character, will form salts and get etched out. Hence we obtained mean phospholipid capturing capacity for TiO_2, Fe_2O_3 and NiO. In line with our discussion in Sect. 4.1.3.4, we have collated all the information in Table 4.2.

Hence, from the data we find that both TiO_2 and Fe_2O_3 can be very good choices for being SPE resins in sample preparation for lipidomics. Let us consider our best performer Fe_2O_3. We injected 20 μl of 1 mg/ml lipid sample. Hence, we have

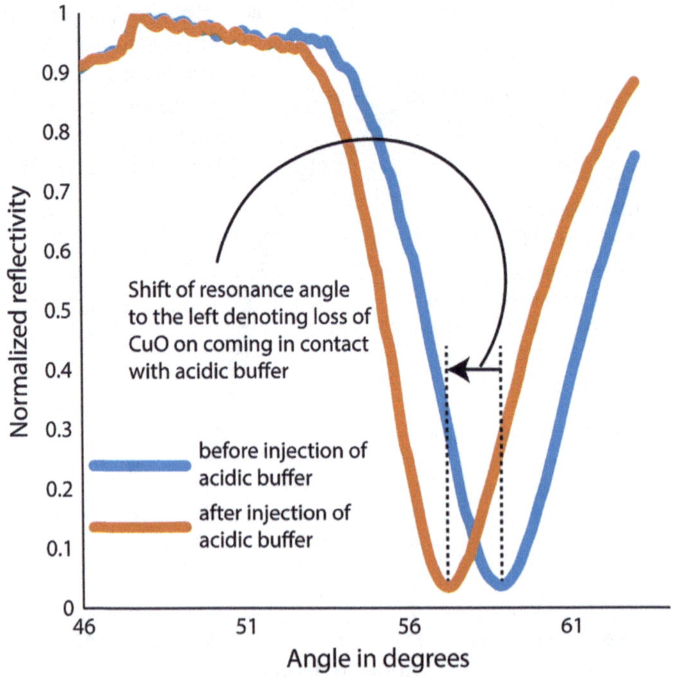

Fig. 4.11 Angular spectrum showing the etching of CuO film when it comes in contact with the 0.1% TFA (acidic buffer) used to inject the phospholipid DMPC into the setup

Table 4.2 Selection of a SPE resin amongst 3d oxides evaluating all of them on parameters discussed in the motivation Sect. 4.1.3.4

Label	Environmentally stable	Forms salts easily in contact with acidic buffer	Phospholipid captured (ng/mm^2)	Cost in SGD (for 5 g of 99.999% purity)
Sc_2O_3	Yes	Yes	N.A.	
TiO_2	Yes	No	1.6415 ± 0.2344	182.50
V_2O_5	Yes	Yes	N.A.	
Cr_2O_3	No	N.A.	N.A.	
MnO	Yes	Yes	N.A.	
Fe_2O_3	Yes	No	1.6520 ± 0.4652	158.50
CoO	No	N.A.	N.A.	
NiO	Yes	No	0.6650 ± 0.3820	109.00
CuO	Yes	Yes	N.A.	
ZnO	Yes	Yes	N.A.	

Costs taken from Sigma-Aldrich, Singapore website

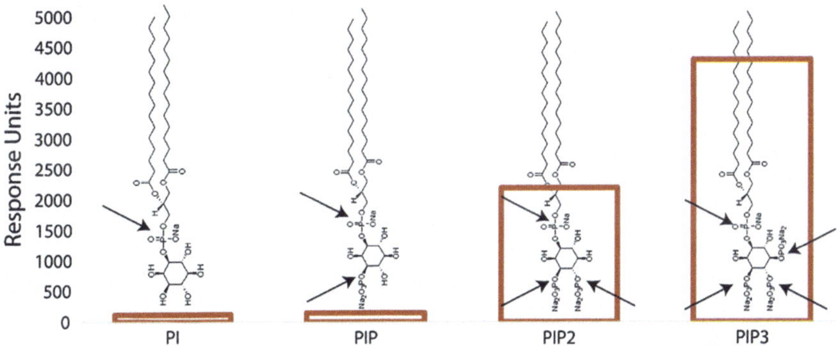

Fig. 4.12 Differences in adhesion based on number of phosphate groups. Phosphate groups have been marked with *black arrow*

injected 20 μg. 1.65 ng is captured per sq. mm. The area in contact is about 10 mm^2. Therefore we have captured 16.5 ng which is 0.0825%. However, we have to remember this is a flat surface.

Let us put these numbers into perspective. Assuming we have powder of Fe_2O_3 with particle radius of 100 nm, 1 mm^2 is the combined available surface area of about 8×10^6 particles. Just 1 mg of Fe_2O_3 powder of the same particle radius would contain about 5×10^{14} particles—a very attractive proposition indeed!

We then tested the effect of the headgroup on the adhesion of phospholipids to a TiO_2 film. The results are shown in Fig. 4.12. Unlike the previous set of SPR experiments, this experiment was performed on a commercial GE Biacore SPR setup. Reflectivity changes are represented in response units (RU) in commercial SPR systems. Upon looking at adhesion of 4 phospholipids (each with a different number of phosphate groups) to the TiO_2 film, we found that PI and PIP, with 1 and 2 phosphate groups respectively, adhere in almost similar quantities to the surface. However, PIP_2 with 3 phosphate groups and PIP_3 with 4 phosphate groups have almost 20 and 40 times the adhesion respectively. Same concentrations of all lipids were used. We postulate that the high adhesion shown by PIP_2 and PIP_3 is a cumulative effect of two factors. Since phospholipids bind to TiO_2 through the phosphate group, more phosphate groups on a phospholipid molecule would mean more affinity for TiO_2. Moreover, as the affinity is higher, the molecules would probably arrange into such assemblies which shadow less of the TiO_2, leaving more of the film exposed for capturing more phospholipid molecules. We can speculate that for PI and PIP, the lipid molecule lies flat on the surface, in the process causing a shadowing of a significant area of TiO_2 by the carbon backbone of the molecule. However for PIP_2 and PIP_3, Only the inositol rings remain on the surface, pinned to the film by the phosphate groups. The hydrocarbon tail shoots out and away from the surface, casting no shadow and making more TiO_2 area available for further binding events.

It was also found that the basic buffer used to unbind PI and PIP, could not elute PIP_2 and PIP_3 because of their strong adhesion to TiO_2. This points to the

Table 4.3 Comparison of various methods of evaluating phospholipid adhesion to SPE resins

Name of method	Control over contact area	Control over contact time	Detection of resin etching	Precise determination of lipid species captured	Compatible with existing mass spectrometry workflow
MALDI	Yes	No	No	No	Yes
μcolumn	No	No	No	No	Yes
Batch processing	No	Yes	No	No	Yes
SPR	Yes	Yes	Yes	Yes	No

possibility of developing a sequential elution protocol, where basic buffers of increasing strength are used to sequentially elute phospholipids beginning with ones which have low adhesion and ending in ones which have highest adhesion.

4.3 Conclusion

We have demonstrated that metal oxides can indeed act as efficient SPE resins in sample preparation for lipidomics. It will not be an overstatement to state that these SPE resins can also be applied to phosphopeptide enrichment. The SPR based technique to accurately quantify phospholipid adhesion has made it possible to compare between various suitable metal oxides and obtain reproducible data dependent only upon the chemistry of interaction. This method can also be used to test the effect of the headgroup upon its affinity for metal oxide surfaces. A summarizing table containing a simple comparison of the various protocols used to measure phospholipid adhesion (Table 4.3).

We have also shown the possibility of applying the knowledge gained through this experiments towards designing separation devices in two formats—μcolumns and planar columns for on target enrichment.

In the next chapter, we will study proteins, a class of biomolecules much larger and complex than lipids. Having used the SPR technique to study lipid interactions with fabricated surfaces, we apply an advanced form of SPR towards quantifying protein interactions with inorganic surfaces.

4.4 Experimental Details

4.4.1 Fabrication of Modified MALDI Plates

Metal oxide thin films were deposited on the commercially obtained MALDI plates from AbSciex. A Neocera 4 inch PLD (Pulsed laser deposition) system was used for this purpose. The standard MALDI plate is 12 cm × 8 cm in size. Hence it was cut into 3 equal pieces 4 cm × 8 cm each. After the cut was made, the plates were cleaned by sonicating them for 10 min in acetone, methanol and DI water successively and in that order. A 248 nm KrF excimer laser was used to deposit films of TiO_2, ZrO_2, Y_2O_3, $LaAlO_3$ (LAO), $SrTiO_3$ (STO), $BaTiO_3$ (BTO). Identical conditions for used for the growth of all the oxides: room temperature, oxygen pressure 3.00 mTorr, laser frequency 10 Hz, 40,000 shots. Film thicknesses were in the range of 240–290 nm.

4.4.2 Lipid Adhesion Studies with Modified MALDI Plates

The following lipids were procured commercially from Avanti Polar Lipids (Alabaster, AL, USA)

1. 1,2-dimyristoyl-sn-glycero-3-phosphocholine (DMPC)
2. N-lauroyl-D-erythro-sphingosylphosphorylcholine (SM)
3. D-glucosyl-ß-1,1′-N-octanoyl-D-erythro-sphingosine (GluCer)
4. N-octanoyl-D-erythro-sphingosine (Cer 534)
5. N-heptadecanoyl-D-erythro-sphingosine (Cer 574)

The steps are outlined below.

- Make up a standard mixture of 1 mg/mL of each lipids in methanol consisting of DMPC, SM, GluCer, Cer 534 and Cer 574
- Dilute the standard mixture to 50 μg/mL with Methanol:H_2O 70:30
- Spot 0.5 μL of the standard mixture in 3 replicates and dry
- Wash with 200 μL of MeCN + 0.1%TFA (trifluoroacetic acid) on the spots thrice
- Dry the spots and spot 0.5 μL of the CLCCA (4-Chloro-α-cyanocinnamic acid) +2,5DHB (Dihydroxybenzoic Acid) matrix onto the spots and dry
- For control, 0.5 μL of the standard mixture in 3 replicates were spot and dry, followed by 0.5 μL of the CLCCA +2,5DHB matrix onto the spots and dry. These are the spots without wash.

4.4.3 Lipid Adhesion Studies with μcolumn Approach

The detailed procedure for the μcolumn setup is given below

- Make up a standard mixture of 1 mg/mL of each lipids in methanol consisting of DMPC, SM, GluCer, Cer.
- Dilute the standard mixture to 50 μg/mL with Methanol:H_2O 70:30.
- Suspend 30 mg of metal oxide powder in 1 ml of acetone. Sonicate for 5 min. Centrifuge the suspension for 14,000 RPM for 10 min. Discard the supernatant

and pipette in 1 ml of fresh acetone. Repeat the process three times. Finally pipette in 1 ml of acetone and sonicate for 5 min. This becomes the suspension used to pack metal oxide powders in the μcolumn.

- Take a 10 μL pipette tip and block the narrow end of it using a stamped out piece of 3 M Empore C8 disk.
- Take 100 μL of the suspension and inject it into the pipette tip. Use a syringe to push out the excess acetone. This is the modified μcolumn.
- Equilibrate the μColumn with 20 μL of loading buffer (Ethanol + 0.1% TFA) for 5 times.
- Load 3 μL of stardard mixture (50 μg/mL) with 17 μL of loading buffer. Push the syringe and collect this as flow through.
- Wash the μColumn with 20 μL of loading buffer for 5 times and collect this in the same tube as flow through.
- Elute the bound lipids with 20 μL of elution buffer [Methanol:chloroform:TEA (Triethylamine) (45:45:10)] for 6 times and collect these as elute.
- Dry the samples in Speed Vac for 30 min.
- Resuspend the dried samples in 10 μL of methanol:chloroform (1:1).
- Analyse the samples in Flash Quant.

4.4.4 Lipid Adhesion Studies in Batch Processing Approach

The procedure is described below

- Prepare a 1 ml standard mixture of lipids by mixing lipid standards of DMPC, SM, GluCer and Cer 17. The final concentrations of these four lipids will be 0.05, 0.05, 0.4 and 0.4 mg/ml respectively.
- Suspend 30 mg of metal oxide powder in 1 ml of ethanol. Sonicate for 5 min. Centrifuge the suspension for 14,000 RPM for 10 min. Discard the supernatant and pipette in 1 ml of fresh ethanol. Repeat the process three times. Finally pipette in 1 ml of ethanol and sonicate for 5 min. This becomes the stock suspension.
- Take 60 μL of stock suspension, centrifuge at 14,000 RPM for 10 min and discard the ethanol. Pipette in 200 μL 0.1% TFA solution in ethanol and sonicate.
- Add 1 μL of standard mixture and shake for 5 min.
- Centrifuge at 14,000 RPM for 10 min and collect the supernatant. Label it "flowthrough."
- Add 200 μL of 0.1% TFA and shake for 5 min. Centrifuge at 14,000 RPM for 10 min and collect supernatant. Add it to the flowthrough collected from previous step. Repeat twice.
- Add 200 μL of basic elution buffer [Methanol:chloroform:TEA (Triethylamine) (45:45:10)]. Shake for 5 min. Centrifuge at 14,000 RPM for 10 min and collect the supernatant. Label it "elution."

- Add 200 μL of basic elution buffer [Methanol:chloroform:TEA (Triethylamine) (45:45:10)] and shake for 5 min. Centrifuge at 14,000 RPM for 10 min and collect supernatant. Add it to the elution collected from previous step. Repeat twice.
- Dry in speed vac for 30 min and resuspend in in 60% water, 40% acetonitrile, 10 mm ammonium formate.
- Analyse samples in ESI-MS.

4.4.5 *Surface Plasmon Resonance*

The chips were fabricated using the PLD technique. The growth conditions were 300 °C temperature, oxygen pressure 3.00 mTorr, laser frequency 1 Hz. The uniformity of the film over the 10 mm × 10 mm area and the elemental composition was checked and the thicknesses were calibrated using RBS. Finally, 3 nm thin films of oxides were deposited on the gold chip.

The running buffer was 10% ethanol. The acidic buffer was 0.1% TFA in 70% ethanol. DMPC was dissolved in the acidic buffer to a concentration of 0.2 mg/ml. The basic eluting buffer is 0.1 N NH_4OH in 70% ethanol. DMPC was injected in the following condition: 160 μl of 0.1% TFA in 70% ethanol was added to 40 μl of 1 mg/ml DMPC. The flow rate was 1 μl/s or 60 μl/min. At this high flow rate, the interaction is not limited by mass transfer. The injection time for both DMPC and the basic eluting buffer is 180 s.

After the data is obtained, WINSPALL software package is used to fit the curves in order to find the change in refractive index ($\Delta n = n_{\text{after binding}} - n_{\text{before binding}}$) and the thickness (d) of the captured layer of phospholipids. The mass on the surface of an oxide thin film is calculated by the following formula.

$$Total\ mass = \Delta n \times \frac{d}{\frac{\partial n}{\partial C}}$$

Δn and d are obtained from the fitting curve. The quantity $\frac{\partial n}{\partial C}$ is change in refractive index per unit change in concentration and is taken to be 0.2 for all proteins, lipids and polymer systems.

References

1. Rozen, S., et al.: Metabolomic analysis and signatures in motor neuron disease. Metabolomics **1**(2), 101–108 (2005)
2. Paige, L.A., et al.: A preliminary metabolomic analysis of older adults with and without depression. Int. J. Geriatr. Psychiatry **22**(5), 418–423 (2007)
3. Holmes, E., et al.: Metabolic profiling of CSF: evidence that early intervention may impact on disease progression and outcome in schizophrenia. PLoS Med **3**(8), e327 (2006)

4. Kaddurah-Daouk, R.: Metabolic profiling of patients with schizophrenia. PLoS Med **3**(8), e363 (2006)
5. Kaddurah-Daouk, R., et al.: Metabolomic mapping of atypical antipsychotic effects in schizophrenia. Mol. Psychiatry **12**(10), 934–945 (2007)
6. Han, X., et al.: Substantial sulfatide deficiency and ceramide elevation in very early Alzheimer's disease: potential role in disease pathogenesis. J. Neurochem. **82**(4), 809–818 (2002)
7. Brindle, J.T., et al.: Rapid and noninvasive diagnosis of the presence and severity of coronary heart disease using 1H-NMR-based metabonomics. Nat. Med. **8**(12), 1439–1445 (2002)
8. Sabatine, M.S., et al.: Metabolomic identification of novel biomarkers of myocardial ischemia. Circulation **112**(25), 3868–3875 (2005)
9. Brindle, J.T., et al.: Application of chemometrics to 1 H NMR spectroscopic data to investigate a relationship between human serum metabolic profiles and hypertension. Analyst **128**(1), 32–36 (2003)
10. Dunne, V.G., et al.: Metabolites from cerebrospinal fluid in aneurysmal subarachnoid haemorrhage correlate with vasospasm and clinical outcome: a pattern-recognition 1H NMR study. NMR Biomed. **18**(1), 24–33 (2005)
11. Kenny, L.C., et al.: Novel biomarkers for pre-eclampsia detected using metabolomics and machine learning. Metabolomics **1**(3), 227–234 (2005)
12. Van der Greef, J., et al.: The art and practice of systems biology in medicine: mapping patterns of relationships. J. Proteome Res. **6**(4), 1540–1559 (2007)
13. Wang, C., et al.: Plasma phospholipid metabolic profiling and biomarkers of type 2 diabetes mellitus based on high-performance liquid chromatography/electrospray mass spectrometry and multivariate statistical analysis. Anal. Chem. **77**(13), 4108–4116 (2005)
14. Yuan, K., et al.: A GC-based metabonomics investigation of type 2 diabetes by organic acids metabolic profile. J. Chromatogr. B **850**(1), 236–240 (2007)
15. Yang, J., et al.: Diagnosis of liver cancer using HPLC-based metabonomics avoiding false-positive result from hepatitis and hepatocirrhosis diseases. J. Chromatogr. B **813**(1), 59–65 (2004)
16. Odunsi, K., et al.: Detection of epithelial ovarian cancer using 1H-NMR-based metabonomics. Int. J. Cancer **113**(5), 782–788 (2005)
17. Fan, X., Bai, J., Shen, P.: Diagnosis of breast cancer using HPLC metabonomics fingerprints coupled with computational methods. In: Conference proceedings:... Annual International Conference of the IEEE Engineering in Medicine and Biology Society. IEEE Engineering in Medicine and Biology Society. Annual Conference, 2004
18. Underwood, B.R., et al.: Huntington disease patients and transgenic mice have similar pro-catabolic serum metabolite profiles. Brain **129**(4), 877–886 (2006)
19. Fisher-Wilson, J.: Long-suffering lipids gain respect: technical advances and enhanced understanding of lipid biology fuel a trend toward lipidomics. Sci. **17**, 5 (2003)
20. Lagarde, M., et al.: Lipidomics is emerging. Biochim. Biophys. Acta **1634**(3), 61 (2003)
21. Duzgunes, N., et al.: Proton-induced fusion of oleic acid-phosphatidylethanolamine liposomes. Biochemistry **24**(13), 3091–3098 (1985)
22. Pernet, F., Pelletier, C.J., Milley, J.: Comparison of three solid-phase extraction methods for fatty acid analysis of lipid fractions in tissues of marine bivalves. J. Chromatogr. A **1137**(2), 127–137 (2006)
23. Avila, V.L., et al.: Cleanup of environmental sample extracts using Florisil solid-phase extraction cartridges. J. Chromatogr. Sci. **27**(5), 209–215 (1989)
24. Wei, W., et al.: Elimination of the interference from nitrate ions on oxalic acid in RP-HPLC by solid-phase extraction with nanosized hydroxyapatite. J. Liq. Chromatogr. Relat. Technol. **32**(1), 106–124 (2008)
25. Gerdes, J.C., et al.: Kit comprising solid phase matrix made of aluminum oxide, titanium oxide (Ti_2O_3) and/or modified zirconium dioxide (ZrO_2) for identifying and extracting viral nucleotide sequences. Google Patents, 2006

26. Leitner, A.: Phosphopeptide enrichment using metal oxide affinity chromatography. TrAC Trends Anal. Chem. **29**(2), 177–185 (2010)
27. Lemeer, S., et al.: Online automated in vivo zebrafish phosphoproteomics: from large-scale analysis down to a single embryo. J. Proteome Res. **7**(4), 1555–1564 (2008)
28. Mohammed, S., et al.: Chip-based enrichment and NanoLC—MS/MS analysis of phosphopeptides from whole lysates. J. Proteome Res. **7**(4), 1565–1571 (2008)
29. Bruce, S.J., et al.: Investigation of human blood plasma sample preparation for performing metabolomics using ultrahigh performance liquid chromatography/mass spectrometry. Anal. Chem. **81**(9), 3285–3296 (2009)
30. Moco, S., et al.: Metabolomics technologies and metabolite identification. TrAC Trends Anal. Chem. **26**(9), 855–866 (2007)
31. Teahan, O., et al.: Impact of analytical bias in metabonomic studies of human blood serum and plasma. Anal. Chem. **78**(13), 4307–4318 (2006)
32. Ikeguchi, Y., Nakamura, H.: Selective enrichment of phospholipids by titania. Anal. Sci. **16**(5), 541–543 (2000)
33. Calvano, C., Jensen, O., Zambonin, C.: Selective extraction of phospholipids from dairy products by micro-solid phase extraction based on titanium dioxide microcolumns followed by MALDI-TOF-MS analysis. Anal. Bioanal. Chem. **394**(5), 1453–1461 (2009)
34. Shen, Q., et al.: Lipidomic study of olive fruit and oil using TiO_2 nanoparticle based matrix solid-phase dispersion and MALDI-TOF/MS. Food Res. Int. **54**(2), 2054–2061 (2013)
35. Polson, C., et al.: Optimization of protein precipitation based upon effectiveness of protein removal and ionization effect in liquid chromatography–tandem mass spectrometry. J. Chromatogr. B **785**(2), 263–275 (2003)
36. Vuckovic, D., Pawliszyn, J.: Systematic evaluation of solid-phase microextraction coatings for untargeted metabolomic profiling of biological fluids by liquid chromatography—mass spectrometry. Anal. Chem. **83**(6), 1944–1954 (2011)
37. Quehenberger, O., et al.: Lipidomics reveals a remarkable diversity of lipids in human plasma. J. lipid Res. **51**(11), 3299–3305 (2010)

Chapter 5
Quantifying Interactions Between Serum Proteins and Gold Nanoparticles

Abstract Understanding the interactions of serum proteins to surfaces are of paramount importance to the domain of nano-drug delivery systems. At the nano-bio interface, human plasma differentially interacts with engineered nanomaterials through the creation of protein coronas, which in turn become primary determinants of both the pharmacokinetics and pharmacodynamics of the circulating nanoparticles. Here, for the first time, the specific binding kinetics of the four major corona forming proteins (human serum albumin, fibrinogen, ApoA1, and polyclonal IgG) are determined for gold nanoparticles (AuNPs). Using a multiplexed surface plasmonic assay, highly reproducible measurements of on rate (k_{on}), off rate (k_{off}), and dissociation constant (K_D), in addition to relative amounts of protein binding, were obtained. Dramatic differences in k_{on} for individual components were shown as primary determinants of protein affinities, with k_{on} ranging over nearly two orders of magnitude for the proteins studied, while k_{off} remained within a factor of two for the set. The effect of polyethylene glycol (PEG) modification on plasma component binding was also studied and the effect of PEG length on human serum interaction was characterized through systematic screening of PEG molecular weight (2–30 K). The effect of nanoparticle modification on particle targeting was also characterized through study of a hybrid AuNP system.

5.1 Introduction

5.1.1 Overview

In the last chapter, we looked at how we could use the SPR technique could be used to accurately quantify phospholipid-metal oxide interactions. A more complex group of biomolecules, perhaps the most important, are proteins.

The "nano-bio" interface has been a domain of high interest in recent years. As nanoparticles find increasing use in both diagnostic and therapeutic roles, the interface between engineered materials and complex human biology is experiencing an expanded focus due to its critical implications for nanomedicine and

A. Patra, *Quantifying Interactions of Biomolecules with Inorganic Surfaces*,
Springer Theses, DOI 10.1007/978-3-319-30728-2_5

nanotoxicity [1–6]. Upon gaining access to the circulation, nanoparticles undergo an association with plasma components, forming a non-covalent protein corona around their surface. It is now known that this protein corona is dynamic in nature and that its identity changes over time [7, 8]. The more abundant proteins bind to the exposed surface of the NP upon contact. Thereafter there is a dynamic competition between the species already present on the surface and additional components which have slower association rates, but ultimately higher affinity for the surface [9, 10]. It is also well understood that the chemical nature, charge, shape and size of the NPs all play significant roles in the formation of the corona [11, 12].

5.1.2 Existing Methods Used to Study Corona Formation and Their Limitations

The formation of the biomolecular corona plays a critical role in the fate of the NP complex [13, 14]. Hence, a platform to characterize the formation, evolution and kinetics of corona on various nanomaterials becomes crucial for applications such as in vitro diagnostic sensors, in vivo diagnostic probes or drug delivery technologies [15]. A commonly used method to study the protein corona formed around NPs is the enzymatic digestion of bound protein followed by polyacrylamide gel electrophoresis and/or mass spectroscopy [16–20]. While this method gives insight to the distribution of bound entities, it is not without limitations, i.e., taking significant manual effort with a large number of sequential experimental steps and typically 8–12 days [16] for a complete workflow. Dynamic Light Scattering (DLS) [21], Differential Centrifugal Sedimentation (DCS) [21, 22] and Size Exclusion Chromatography (SEC) [23] are additional, useful techniques that characterize the diffusional radius of nanoparticles and their bound coronas, however kinetic analyses are largely precluded due to technical limitations. As DLS is limited to colloidally stable NPs with a narrow size distribution (typically larger than 10 nm) and often has difficulty resolving unbound from bound corona components, Fluorescence Correlation Spectroscopy (FCS) [24, 25] has since emerged as an alternative to DLS, but it requires NPs to be first modified with a fluorescent tag [26]. The affinity of proteins for the NPs may not be the same as their fluorescently labelled forms. Moreover, FCS is sensitive to the fluorescence of protein-NP complexes only if the autofluorescence from unbound proteins can be neglected. An additional drawback with DLS and FCS is their inability to provide meaningful data for anisotropic shapes [17]. Furthermore, if the change in hydrodynamic radius of the protein-NP complex after adsorption is not significantly large, neither DLS nor FCS can feasibly resolve the change [27]. Hence the methods are usually used to investigate small NPs (but much larger than proteins themselves). There is a need for a method which can return kinetic data from real-time interactions, be sufficiently adaptable to work for a variety of sizes, shapes and chemical compositions and enable high-throughput data collection with high precision within reasonable time frames.

5.1.3 Need for a New Method and Previously Reported Work

Surface plasmon resonance (SPR) is a technique that uses disruptions to surface plasmons, i.e. delocalized electrons on, typically, a gold surface, to study binding reactions with time resolved high fidelity measurements. Cedervall et al. [23] used SPR to observe dilute plasma modification of a biopolymeric NP, noting wide differences in disassociation rates as a function of hydrophobic content of the particle. A more recent study examined the effects of protein components on a SPR surface as they effected liposome absorption [28]. Although both studies demonstrated the utility of SPR in time resolved NP recordings, neither study was able to report affinity constants or kinetic determinants of NP modification by individual plasma components.

In this study, we have chosen AuNPs as our model system because of their ease of synthesis, characterization and relevance/use in targeted delivery. We have immobilized native AuNPs on the surface without additional functionalization, thus enabling the study of serum proteins with the native surface of the AuNPs. Beyond studying the interactions of chosen serum proteins with native AuNPs, we used our protocol to study both PEG modification effects on blood components and hybrid targeted AuNPs interactions with serum.

To better understand the kinetic determinants of plasma component affinities to AuNPs, in addition to their relative adhesion efficiencies, we developed a multiplexed SPR assay using a commercially available (BioRad XPR36 SPR platform) reaction surface, thus enabling the extensive control measurements needed for subtraction of non-specific and background binding. Through a layer-by-layer modification strategy, we could monitor construction of modified AuNP's and then analyze them to obtain kinetic rate constants and relative adhesion amounts for specific plasma components, with experimental runs lasting no more than 3 h.

5.2 Results and Discussion

5.2.1 Immobilization of Gold NPs

The gold surface of the GLC^{TM} chip from BioRad is supplied with a modified alginate polymer which provides a net negative charge and active sites for ligand immobilization. The alginate polymer layer also prevents non-specific binding to the chip surface—a common artefact in SPR experiments. Sulfo-NHS (N-hydroxysulfosuccinimide) enables control and modification of carbodiimide crosslinking reactions involving activation of carboxylates (–COOH) for conjugation with primary amines (–NH_2). Derivatives are synthesized by mixing sulfo-NHS with a carboxyl-containing molecule and a dehydrating agent which in our case is 1-ethyl-3-(3-dimethylaminopropyl) carbodiimide (EDAC). Hence, the linker amino-EG6-undecanethiol hydrochloride, is immobilized onto the chip through amide coupling. The corresponding change in SPR signal can be seen from time 0 to 600 s in the sensorgram showed in Fig. 5.1a. The –SH end of the linker is now free and the AuNPs are immobilized on the surface of the chip through these free –SH groups. It is possible to monitor the immobilization in real time as seen from t = 950 to 1700 s in Fig. 5.1a. The distribution and density of immobilized AuNPs was assessed directly using SEM as seen in Fig. 5.1b, allowing us to correlate SPR response with the final density of AuNP coating.

Gold NPs used in this study are citrate-capped. The citrate capping provides long term stability against charge imbalance induced aggregation. Citrate weakly associates with the nanoparticle surface and is readily replaced by thiols. Hence there is no effect of citrate capping on the capture by the chemical linker. One advantage of being able to observe the systematic deposition and modification of nanoparticles in preparation of a downstream experiment is to ensure regularity of the substrate, i.e. differences in reactivity of linkage chemistries, particle formulations, etc. can be seen as aberrant profiles in the assembly stage, as visualized by real time SPR. Hence the method allows for step checks to enhance reproducibility across multiple runs. We would also like to point out the platform nature of our protocol. By changing the linker molecule, it is potentially possible to immobilize additional classes of NP through diverse attachment chemistries and study its interactions, thereafter, with relevant proteins. We also would like to state that the size of the nanoparticle is 17 nm and its peak SPR wavelength is about 520 nm. The laser used in our study has a wavelength of 632.8 nm and hence there is no plasmonic coupling. At this size of the nanoparticle, there is absolutely no effect of gold on the evanescent wave per se. Hence, the immobilized gold NP could very well be any other NP.

Upon analyzing the SEM image shown in Fig. 5.2a, we find that there are 1584 AuNPs captured in an area of 1,987,712 nm^2 (1696 nm × 1172 nm).

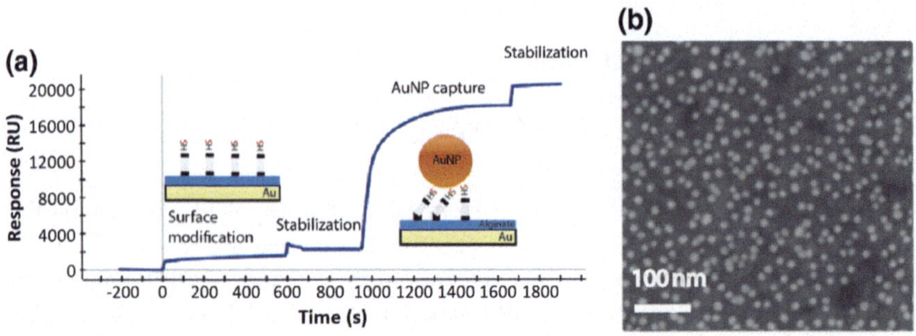

Fig. 5.1 **a** SPR sensorgram of surface modification using a hetero-bifunctional linker and subsequently, the immobilization of AuNPs. **b** SEM image of the sensor surface after AuNP injection

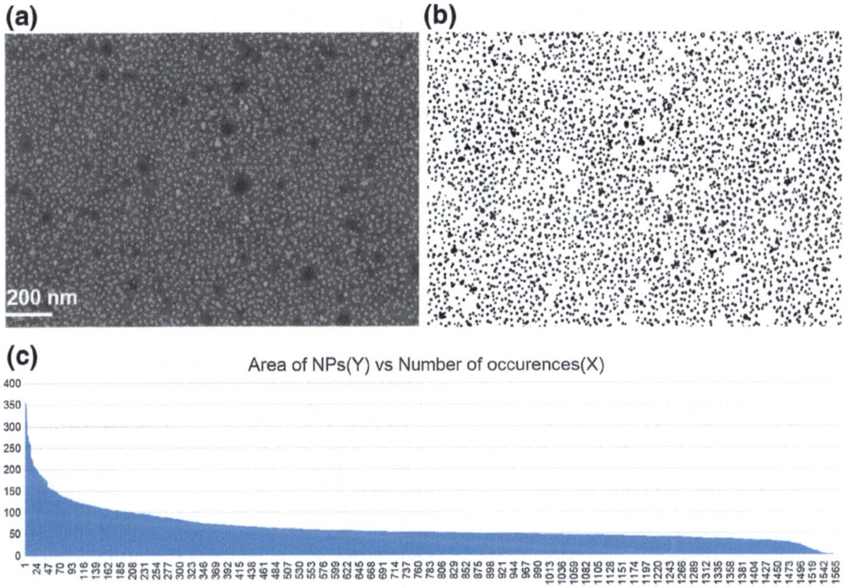

Fig. 5.2 **a** SEM image of the chip surface after immobilization of AuNP. **b** The same image, processed with Image J to enable counting of AuNPs. **c** Distribution of particles on the surface. It can be seen that there is almost no agglomeration and the AuNPs are well dispersed

5.2.2 Studying Interactions of AuNPs with Proteins

One of the major advantages of multiplexing data acquisition using SPR is reproducibility. Owing to the availability of 6 × 6 microfluidic channels leading into 36 interaction spots that can be monitored simultaneously, it is possible to obtain 36 sensorgrams representing 36 interactions in a single experimental run. We designed our experiments to have unambiguous co-temporal controls to compare the experimental data to. These are set out in detail in the caption of Fig. 5.3 where a schematic representation has also been provided.

The four most common serum proteins found in nanoparticle coronas: human serum albumin (HSA), fibrinogen, immunoglobulin G (IgG) and apolipoprotein A1 (ApoA1), were selected for analysis [29–31]. After the AuNPs were immobilized and stabilized on the chip, 3 different concentrations of the 4 chosen proteins were flushed into the chip to observe the interactions. The association rate constant (i.e., k_{on} or on-rate), the dissociation rate constant (i.e., k_{off} or off-rate) and the equilibrium dissociation constant (K_D) were determined through first order kinetic curve fitting (Table 5.1).

The fibrinogen primary data are shown in Fig. 5.3a–c. Sensograms for the three remaining proteins can be found in the Appendix Figs. A.1, A.2 and A.3. The SPR surface has a very thin coating of alginate hydrogel to separate the gold surface from the tethered nanoparticle of interest. To demonstrate lack of interaction of the

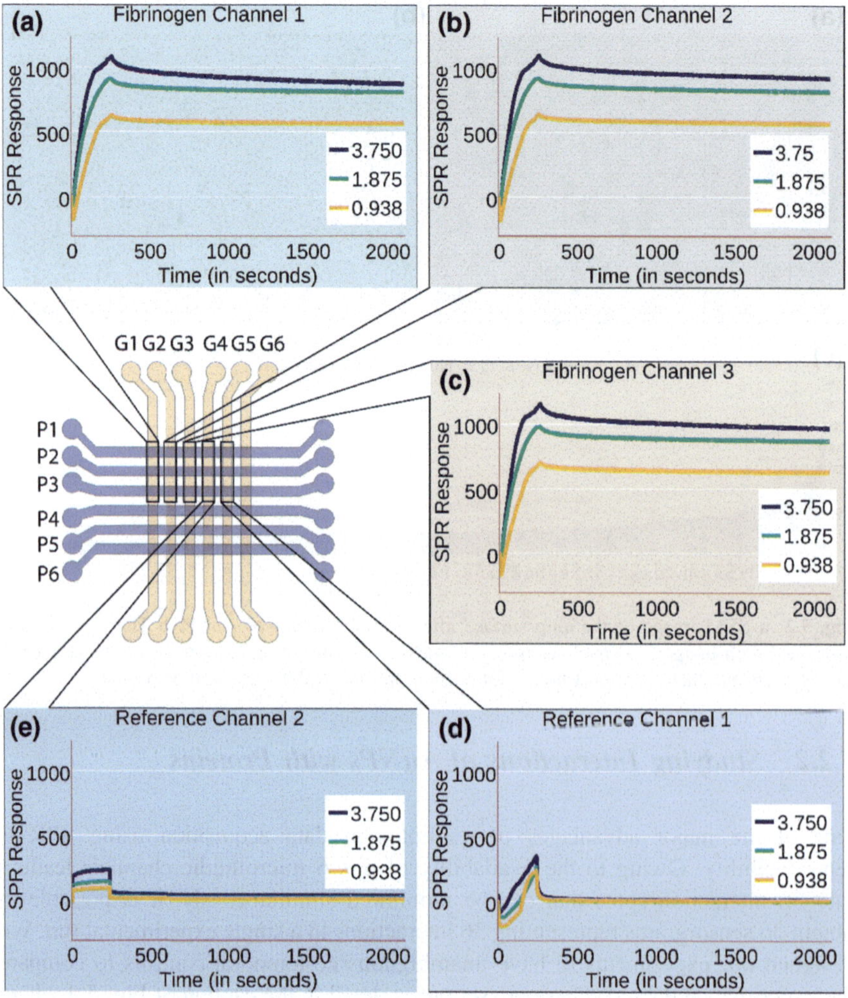

Fig. 5.3 The microfluidic channel setup allowing multiplexed data collection is shown. AuNPs are immobilized in 3 channels G1, G2 and G3 using EDC/S-NHS coupling as described. Two control channels (G4 and G5) are employed. G4 channel is modified with all the linker chemistry used to capture AuNPs (but AuNPs are not injected). G5 channel is in as received condition (alginate layer with inactive –COOH groups). Fibrinogen in three concentrations (3.750, 1.775 and 0.938 μM) is injected along channels P1, P2 and P3 respectively. The three colored lines (*blue*, *green*, *yellow*) in each panel correspond to 3 different concentrations of fibrinogen used in the experiment (legend in *white box* in each panel) **a**, **b** and **c** Sensorgrams showing the interaction of fibrinogen with bare AuNPs available in channels G1, G2 and G3. **d** Sensorgram resulting from the interaction of fibrinogen with the chemically active lane without the AuNPs. **e** Sensorgram showing the interaction of fibrinogen with the unmodified surface (alginate polymer layer)

Table 5.1 Kinetic rate constants and equilibrium dissociation constants for four chosen serum proteins as calculated from fitting curves to the experimentally obtained SPR responses

Label	k_{on} (M^{-1} s^{-1})	k_{off} (s^{-1})	K_D (μM)
HSA	466.860 ± 252.10	0.00184 ± 0.00033	04.93 ± 2.41
Fibrinogen	3857.330 ± 1031.08	0.00205 ± 0.00043	00.53 ± 0.08
I_gG	232.47 ± 24.19	0.00233 ± 0.00068	10.13 ± 3.28
ApoA1	18634.00 ± 9590.50	0.00163 ± 0.00007	00.12 ± 0.07

gold surface with the AuNP's, we employed a negative control lane wherein AuNP is flowed above the surface without activated linker chemistry. No interaction was observed in this setting. This is likely due to the separating effect of the thin layer of alginate hydrogel on the surface of the chip. Alginate typically forms a hydrogel with greater than 98% water content, creating a near native surface for AuNP attachment and minimizing surface adsorption (experiments with corona components and no AuNP also showed no interaction). Thus, although SPR has generally been an accepted methodology for component—specific kinetic studies of many types, there may exist differences with AuNP's experiencing full diffusional freedom, i.e. the corona kinetics of freely dissolved particles or colloidal suspensions.

ApoA1 has the lowest K_D of 0.12 ± 0.07 μM, implying that it binds most strongly to the surface of the AuNP. Fibrinogen follows with a K_D of 0.53 ± 0.08 μM. The affinity of HSA affinity for the surface of AuNP is about 10-fold lower than that of fibrinogen, bearing a K_D of 4.93 ± 2.41 μM. IgG is the component with least affinity (amongst the four chosen serum components) bearing a K_D of 10.13 ± 3.28 μM, implying it is about half as active as HSA with regard to adsorption onto the AuNP surface. Tsai et al. [32] have reported K_D values for interaction of BSA with AuNPs using Langmuir adsorption models on data obtained through electrospray-differential mobility analysis (1.96 μM), fluorescence assay (1.61 μM) and attenuated total reflectance Fourier transform infrared spectroscopy (0.6 μM). As can be seen, the values of K_D for HSA-AuNP adsorption as obtained through SPR are within the same order of magnitude as BSA-AuNP adsorption, previously demonstrated through 3 different methods. To the best of our knowledge, kinetic and formal affinity constant data for the other major components interacting with AuNPs have not been previously reported.

We observed remarkable differences in k_{on} as obtained for the four proteins, while k_{off} values were comparable. ApoA1 (highest affinity) binds to the AuNP surface with an association constant almost 100 times higher than IgG (lowest affinity). Overall, on-rates were found to play by far the largest role in the determination of component specific particle affinity for AuNPs.

Studies of absolute distributions of known plasma components are rare, the most relevant prior study may be Mahmoudi et al. [33] which described corona component binding to gold nanorods. This study used normalized spectral counts on LCMS (Liquid chromatography–mass spectrometry) to assign a relative binding amount of HSA and ApoA1 of 5.4 and 14.95, or a ratio of 1:2.8. Comparing values

Table 5.2 Estimated surface coverage of individual proteins in protein corona

Label	K_D (μM)	$K_A = 1/K_D$	Physiological concentration (μM)	K_A (Protein)
HSA	4.93	0.20284	630	127.789
Fibrinogen	0.53	1.886792	7.8	14.71698
I_gG	10.13	0.098717	83	8.193485
ApoA1	0.12	8.333333	44	366.6667

calculated using our kinetic measurements and known plasma concentrations of HSA and ApoA1, we arrive at a very similar ratio of 1:3 (Table 5.2), consistent with the ability to approximate relative contributions to corona formation based on individualized kinetic and affinity measurements.

One significant limitation of the SPR approach is that component-specific kinetic data can only be extracted from sensorgrams using purified reagents. In the setting of complex biofluids, serum, plasma, urine, etc., aggregate rates of modification can be observed, however measurement of component-specific rate constants in this setting is generally not possible.

There are also limitations to kinetic analysis as performed through SPR. Largely due to mass transport limitations, it is difficult to measure k_{on} values faster than about 10^6 M^{-1} s^{-1} accurately [34]. This upper limit is dependent on the size of the analyte. In case of analytes with high molecular mass, faster k_{on} values can be measured. This is because the larger signal produced by a large analyte allows the experiment to be performed at lower ligand densities, and lower ligand densities require lower rates of mass transport. For instance, a higher k_{on} of 10^7 M^{-1} s^{-1} was measured by SPR for the interaction between HIV Tat protein with integrin $\alpha_v\beta_3$ [35]. In case of k_{off}, values slower than 10^{-5} s^{-1} can be difficult to accurately measure, mainly due to issues with instrument drift over long periods of time [34, 36].

5.2.3 *Effect of PEGylation Observed Through SPR*

To evaluate the effect of PEG modification on plasma component binding, we used our multiplexed SPR protocol to establish saturating modification of thiolated PEG onto the AuNP based on the real-time SPR signal. Once surface saturation of PEG was achieved, we were able to demonstrate differential effects on plasma components and PEG length-specific effects on NP modification by serum. Modification of injectable therapeutics with PEG is well known to affect circulation half-life and decrease protein binding [37, 38]. However, more specific effects on individual plasma components is less clear. After immobilizing AuNPs on the chip surface, PEG-2000 was injected. To address the possibility that thiol groups on PEG-2000 may displace the immobilized AuNPs from the assay chip, we verified retention of the AuNPs through SEM images after the experiment was completed. We observed that PEGylation was effective at preventing adhesion of all four serum proteins (Fig. 5.4). The level of signal originating from adhesion of proteins was normalized to the level of signal change observed when the AuNPs were immobilized on the surface. It is seen that PEGylation is about 99% effective in blocking HSA (Fig. 5.4a) while PEGylation is about 94% effective against fibrinogen (Fig. 5.4b). The adsorption of IgG was reduced by about 93% (Fig. 5.4c). PEGylation was found to be very effective against ApoA1 (Fig. 5.4d) adsorption as well, showing a 98% decrease.

It is important to note the significance of adsorption of proteins on NPs functionalized with antifouling agents (in this case PEG). Even though the amount of adsorption is small, if the adsorbing protein triggers cell surface receptors, the uptake machinery of the cell may be activated. In turn, this would lead to internalization of the NP by the reticuloendothelial system (RES) and thereby reduce circulation time. Further, the adsorption of other complement proteins such as C1q, MBL, ficolin and C3b, even at trace amount, would be sufficient to trigger an undesirable complement activation.

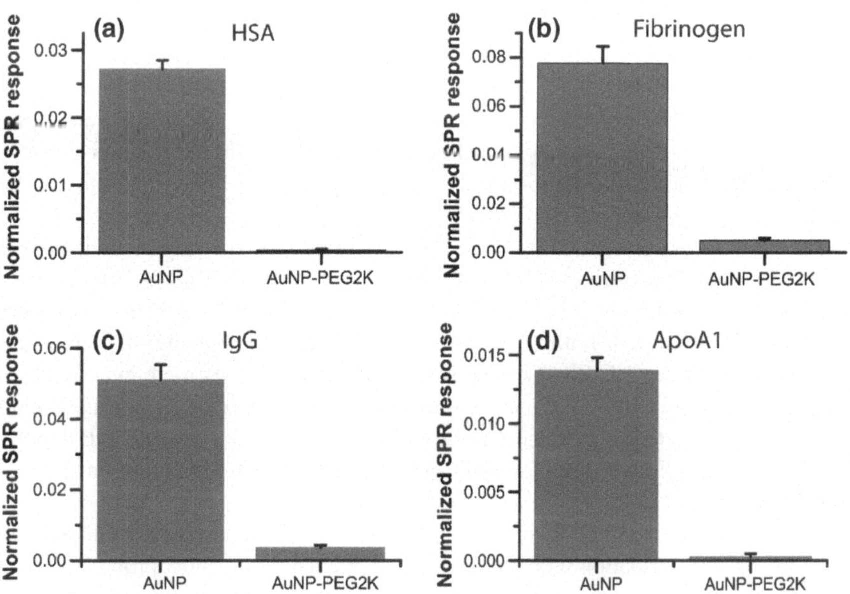

Fig. 5.4 Normalized SPR response for interaction of individual serum proteins (**a** HSA, **b** Fibrinogen, **c** IgG, **d** ApoA1) with AuNPs modified with PEG2K compared to unmodified AuNPs. The absolute response obtained due to adsorption of proteins to modified and unmodified AuNPs is divided by the absolute response caused by the immobilization of AuNPs on the sensor surface. This normalization, hence, takes care of the minor differences in number of AuNPs captured in each lane

5.2.4 *Effect of Varying PEG Chain Length on Resistance to Corona*

After having validated the effect of PEGylation on individual plasma components, we decided to investigate the effect of PEG chain length moieties on the resistance of the functionalized NP to corona formation. To accomplish this study, PEG molecules having average molecular weight (MW) of 2, 5, 10, 20 and 30 K were chosen. After duly immobilizing AuNPs on the chip surface as before, we injected the chosen PEG moieties, each in a separate lane, leaving a control lane with unmodified AuNPs. This lane would serve as a control to compare our findings to. All absolute responses were normalized to the SPR response resulting from the capture of AuNPs in the designated lanes, to allow for cross comparability. To avoid problems due to probable differential affinities of varying PEG chains to the gold surface, every sample of PEG was injected thrice with an aim to saturate the surface of the gold with PEG. This enabled us to get as close to complete coverage of surface as possible so that artefacts due to partial coverage of exposed surface could be minimized.

The nanomedicine community has long been dealing with the "PEG dilemma" [39]. Due to the presence of its long apolar chains, PEG has been shown to reduce protein adsorption. However in recent years, PEGylation has also been shown to strongly inhibit cellular uptake and limit the binding with intended protein targets thereby reducing the potential of drug delivery significantly. Hence, the emerging trend seems to revolve around optimization—such that the functionalization is made to retard protein adsorption but not so much that it limits its ability to be trafficked into cells and deliver its cargo.

Here, we found that PEG 2K, PEG 5K and PEG 10K are very similar in their resistance to corona formation allowing only 14, 10 and 15% adsorption, respectively, as compared with an unmodified surface (Fig. 5.5). PEG 20K and 30K were the least effective anti-fouling moieties of the group allowing 31 and 40% respectively. Hence PEG 2K, 5K and 10K are about 3–4 times more effective than the longer chained ones. Our data is thus consistent with prior reports indicating that shorter PEG chains pack more tightly to the surface of an exposed NP thereby creating a more efficient shield against non-specific protein adsorption on the surface [40].

As the PEG chain length increases, the PEG molecules are understood to adopt a "mushroom" conformation with reduced packing density that allow proteins to bind to available exposed areas on the NP surface [41]. It has been previously suggested that protein adsorption to PEG grafted surfaces is very sensitive to chain density effects and less so to chain length [40, 42]. Although our data is consistent with this hypothesis, further experiments will need to be performed to accurately characterize steric effects of linear and branched PEG molecules. Likewise, there are conflicting reports in literature regarding the effect of the chain length on adsorption. Gref et al. [43] have reported that when polylactic acid (PLA) NPs were functionalized with varying lengths of PEG, the most amount of protein is found adsorbed to PEG 2K.

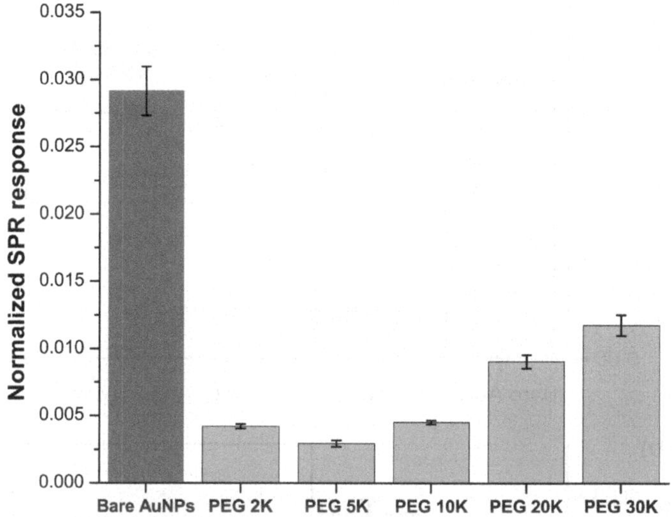

Fig. 5.5 Normalized SPR response for interaction of 20% (W/V) human serum (diluted with PBS buffer) with AuNPs modified with PEG2K, PEG5K, PEG10K, PEG20K and PEG30K compared to unmodified AuNPs. With the exception of PEG 2K and PEG 5K, all sets of data were statistically different from all other sets (p-value $\ll 0.05$)

PEG 5K, 10K, 15K and 20K are all similar in their resistance to adsorption and they adsorb about 40% of protein that is adsorbed by PEG 2K. On the other hand, a more recent report by Pozzi et al. [44] claims that when varying chain lengths of PEG are used to functionalize multicomponent cationic liposomes, PEG 1K adsorbs the most amount of proteins while PEG 2K and PEG 5K both adsorb about 50% of PEG 1K.

We would also like to mention here that in a control experiment designed to determine whether PEG binds to the chip surface. We injected AuNPs in 4 lanes and had an empty lane with all the required chemical linkers to capture the AuNPs, but didn't inject the AuNPs. We found, in the absence of the captured AuNPs, there is no interaction between PEG and the chemically active surface of the SPR chip. The data has been provided in the appendix Fig. A.4.

5.2.5 Observation of Hybrid Modifications Effect on NP Targeting Design

NP technologies engineered for therapeutic applications are often multifunctional, containing materials that target to the site of disease, optimize plasma half-life and provide therapeutic or imaging modalities. When the complexity of NP design is tested in vivo, interactions beyond the four major plasma components may become

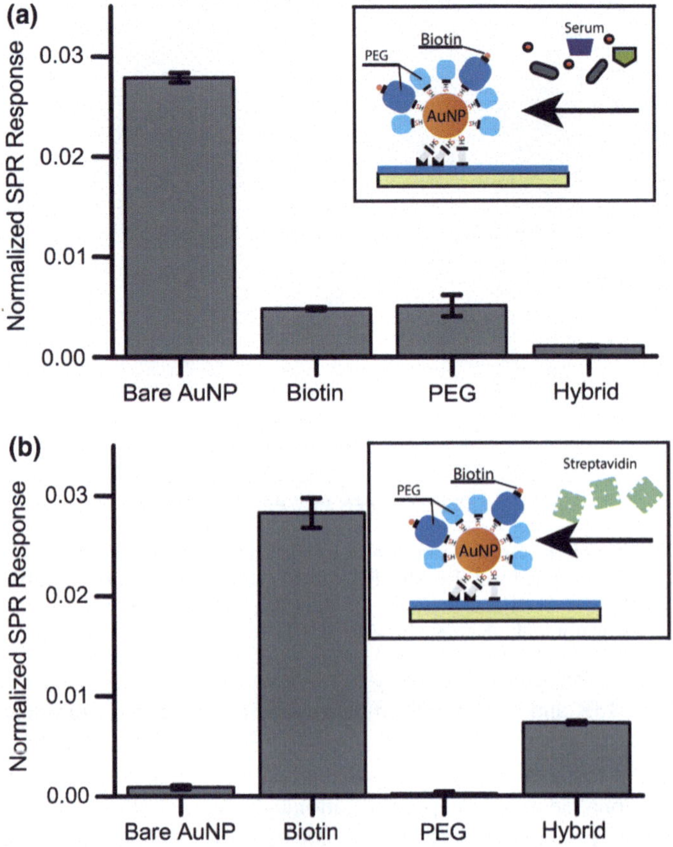

Fig. 5.6 **a** Normalized SPR response for interaction of 20% human serum with AuNPs modified with PEG5K-biotin (referred to as "Biotin"), PEG2K and 1:1 mixture of Biotin:PEG2K (referred to as "Hybrid") compared to unmodified AuNPs. The inset illustrates the schematic of the experiment **b** Normalized SPR response showing the ability of AuNPs with previously described functionalizations to bind intended target streptavidin. The inset illustrates the schematic of the experiment (p-value $\ll 0.05$)

important. In particular, for particles that are functionalized with a specific targeting moiety, the cumulative effects of antifouling modifications may relate directly to binding capacity. Consequently, we synthesized a model NP system (named 'hybrid' in Fig. 5.6) to have a corona retarding moiety—the commonly used SH-PEG2K and a target binding moiety –SH-PEG5K-Biotin (the target being streptavidin). The two controls that were used in this experiment were AuNPs functionalized with only SH-PEG2K and AuNPs functionalized with only SH-PEG5K-biotin (referred to as biotin). It is logical to conclude, in the hybrid configuration, the biotin end of the –SH-PEG5K-Biotin molecule will be accessible above the packing layer of –SH-PEG2K (since PEG5K is longer than PEG2K). Figure 5.6a shows the sensorgram from the interaction of 20% human serum with the functionalized AuNPs. The inset shows the schematic illustration of the interaction. To be able to compare the performance of the NP systems with respect to its resistance to biomolecular fouling and its ability to bind its intended target streptavidin, on each type of NP system, we have normalized the absolute SPR responses resulting from serum interaction and streptavidin binding to the level of gold initially captured in the lane. Compared to unmodified AuNP, all the lanes showed a drastic drop. PEG is well known to resist protein adsorption and hence the surface of the AuNP functionalized with PEG 2K shows 18% of the adsorption signal observed in the bare gold surface AuNPs. Likewise, PEG linked biotin shows about 17% adsorption when compared with unfunctionalized surface. The hybrid system, dramatically, shows only 3% adsorption of when compared with bare AuNP. We hypothesize this might be caused by a favorable interplay between the shielding accorded by the PEG 2K moiety and the PEG 5K section of the PEG5K-biotin moiety. The exact affinity of this hybrid system for serum proteins may perhaps be tailored by playing upon the ratio between SH-PEG and SH-PEG-biotin moieties.

In NP formulations, addition of new components almost always involves a trade-off with other functionalities. We thus studied the trade-off of biotin streptavidin binding in the setting of additional PEG modification, in the highly controlled setting of multiplexed SPR. Figure 5.6b shows the sensorgram for the interaction of streptavidin with these modified AuNPs (schematic in inset). As expected, the unmodified AuNP showed almost no non-specific adsorption of streptavidin (3%). The AuNP with PEG alone showed no response because there was no biotin present. However, when we examined the biotin and hybrid systems, we were able to determine the precise trade-off in terms of streptavidin binding with 26% of streptavidin bound when compared with the original biotin-only particle. That the interaction is truly between biotin-streptavidin is evident from the extremely low off-rate and lack of interaction seen in control lanes. It should also be stated that since the surface of the AuNP in the case of the hybrid is shared by –SH-PEG2K and –SH-PEG5K-biotin, the number of available biotin groups for binding streptavidin will be much lesser than the case where the AuNP is modified with –SH-PEG5K-biotin alone. Hence a part of the reduced response can be attributed to a lesser number of target-binding ligands in the lane.

5.3 Conclusion

We have established a robust protocol which can be used to study interactions of NPs with biomolecular entities of interest. Using multiplexed SPR measurements, including extensive controls, we were able to accurately determine kinetic measurements of plasma component interactions with AuNPs, including wide variation in association constants for individual plasma components, and conduct comparative studies of corona formation. A comparison with the existing techniques has been provided in the Table 5.3.

Table 5.3 Comparison of methods used to study corona formation on nanoparticles

Name of method	Pros	Cons	Error in measurement
DLS, DCS, SEC	1. Quick 2. Nanomaterial independent	1. Can't handle anisotropic shapes 2. Many restrictions on relative size of protein and nanoparticle	High
SDS-PAGE/MS	1. Nanomaterial independent 2. Comprehensive identification of all bound serum proteins	1. Very slow (8–10 days) 2. Large number of manual steps 3. Possibility of user bias	Medium
Multiplexed SPR	1. Nanomaterial independent 2. Can handle anisotropic shapes 3. Real time kinetic parameters of interaction 4. Fully automated—no user bias 5. Can be used to screen drug delivery formulations ex vivo	Size of nanomaterial must be below 100 nm	Low

In principle, the same protocol could be applied to a wide variety of NP materials and formulations as only a single point of attachment is required for the SPR surface and many NPs lie within the theoretical plasmonic sensing range of 100 nm. Although, in principle, the techniques and analysis of tethered polymeric NP's should be very similar to the AuNP's in this study, key differences in operational efficiency will be known only upon additional experimentation. Because test-particles are assembled on-chip, the ability to show saturation kinetics helps to ensure stabilized surface occupancy of modifying agents. The ability to perform multiple assays alongside proper controls using the same reagents on the same chip is of particular value given the multicomponent nature of nanoparticle designs and demonstrates the strength of multiplexed SPR.

5.4 Experimental Details

5.4.1 Synthesis of AuNPs

The NPs were synthesized using the method of Frens et al. [45] 100 mL of 1 mM hydrochloroauric acid ($HAuCl_4$) was heated until it boiled. 15 mL of 38.8 mM trisodium citrate was added to the solution while being vigorously stirred. After being boiled for 15 min, the solution changed color from pale yellow to purple before finally becoming a deep red NP colloid. The citrate-capped NP colloid was washed twice through centrifugation at 10,000 rpm for 15 min and diluted 10X for subsequent experiments. The zeta potential and hydrodynamic diameter (DH) of the NPs were measured at 25 °C using a Zetasizer (Nano ZS, Malvern, UK), and the morphology characterized using transmission electron microscopy (TEM) (JEM-1220, JEOL Ltd., Japan). The concentration of the NPs was determined by optical absorption. The synthesized NPs were kept at 4 °C until further experiments.

5.4.2 Immobilization of AuNPs onto the Chip Surface

A Bio-Rad ProteOn XPR36 instrument (Haifa, Israel) was used in this study. The GLC type sensor chip consists of a glass prism coated with gold and an alginate layer with a low capacity for ligand conjugation. A detailed introduction of experimental protocol for ligand-analyte interactions can be found in literature [46]. The system was first equilibrated with PBS-T buffer (20 mM Na-phosphate, 150 mM NaCl, and 0.05% Tween 20, pH 7.4). The channels of the chip were activated for 5 min with a mixture of EDC (0.2 M) and sulfo-NHS (0.05 M), followed by 10 min injection of 1 mM amino-EG6-undecanethiol (Dojindo Laboratories, Japan) and 5 min injection of ethanolamine-HCl solution. Subsequently, AuNPs were injected onto the modified sensor lane for 6 min with a flow rate of 30 μL/min. The chip is then flushed PBS-T for 30 min to monitor the dissociation of AuNP from the surface.

5.4.3 Studying Interactions of Serum Components with Immobilized AuNPs

Using aforementioned methods, 17 nm AuNPs were captured on the SPR chip surface and stabilized for 30 min. Human serum Albumin (HSA), Human Fibrinogen and Human immunoglobulin G (IgG) and Human Apolipoprotein A1 (ApoA1) were dissolved in PBS and injected into the AuNP modified sensor surface. The injection time was 4 min at a flow rate of 100 μl/min, followed by a dissociation step of 30 min. It has to be noted that the direction of injection of the proteins is orthogonal to the direction in which the AuNPs are injected. Please refer Fig. 5.2 for the layout of the chip. This means that in a single protein lane, we can obtain n interaction sensorgrams ($1 \leq n \leq 6$, depending on how many lanes have been populated with AuNPs).

5.4.4 Obtaining Kinetic Parameters from Curve Fitting and Calculating Surface Coverage

The data obtained from the experiments was exported to and analyzed with Origin software package.

For the dissociation phase,

$$R_d = R_{d0} + A_1 e^{-k_{off} t} \tag{5.1}$$

where R_d is the response at any time t in the dissociation phase, R_{d0} is the response at the start of the as chosen dissociation phase, A is a constant, k_{off} is dissociation constant and t is time in seconds.

In the association phase,

$$R_a = A_2 - A_3 e^{-kt} \tag{5.2}$$

where R_a is the response at any time in the association phase, A_2 and A_3 are constants and t is time in seconds. The parameter k will be used to determine k_{on}, by the formula given below

$$k_{on} C + k_{off} = k \tag{5.3}$$

where k_{off} is dissociation constant calculated from (5.1), k is the parameter obtained from (5.2) and C is concentration of protein in moles/liter.

Finally,

$$K_D = \frac{k_{off}}{k_{on}} \tag{5.4}$$

While fitting curves, care has to be taken to select such portions of the responses which are free from bulk refractive index changes. An example has been provided in Fig. 5.7.

The interaction between the four proteins (A_1, A_2, A_3, A_4) and the AuNPs (B) can be described by the following process,

$$A_1 + B \underset{k_{d2}}{\overset{k_{a1}}{\longleftrightarrow}} A_1 B$$

$$A_2 + B \underset{k_{d2}}{\overset{k_{a2}}{\longleftrightarrow}} A_2 B$$

$$A_3 + B \underset{k_{d3}}{\overset{k_{a3}}{\longleftrightarrow}} A_3 B$$

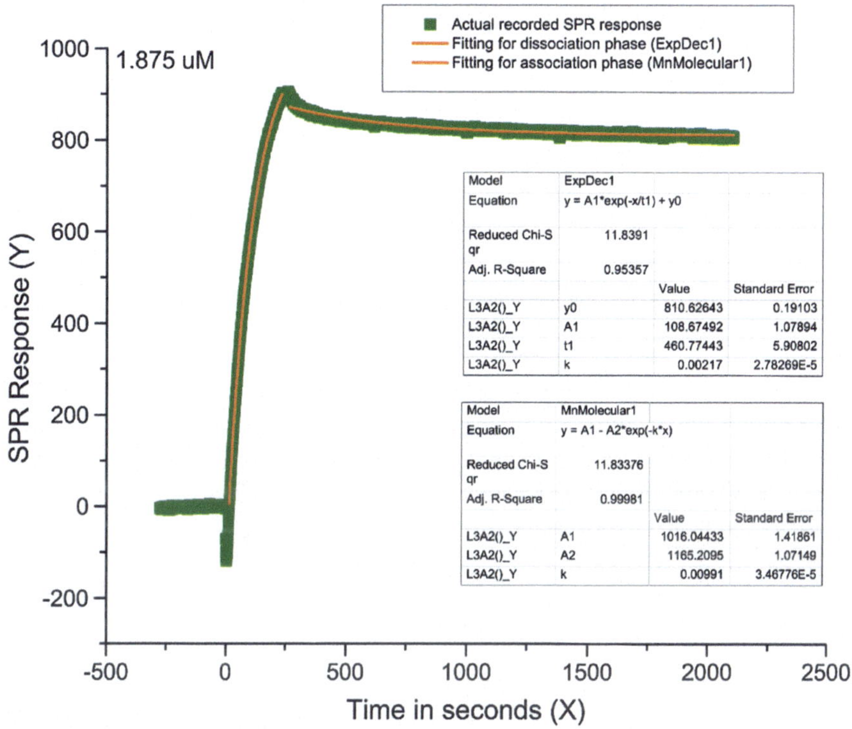

Fig. 5.7 Example of curve fitting on origin software package on experimentally collected data to obtain kinetic parameters

$$A_4 + B \underset{k_{d4}}{\overset{k_{a4}}{\longleftrightarrow}} A_4B$$

Assuming the mass transport is much faster than the surface reaction and the binding kinetics is interaction controlled, we can write the binding rate as

$$\frac{d[AB]}{dt} = k_a[A][B] - k_d[AB], \tag{5.5}$$

where [A] represent the concentrations of A_1, A_2, A_3, A_4. And [B] is the concentration/surface of available AuNPs. As the concentration [A] is much higher than [B], i.e., [A] ≫ [B], the concentration of $[B]$ is the function of time t, which is the difference between the concentration of maximal available concentration of B and the current complex concentration,

$$[B] = [B_{\max}] - [A_1B] - [A_2B] - [A_3B] - [A_4B]. \tag{5.6}$$

Upon dynamic equilibrium, the net effect of the association and dissociation process is zero, i.e.

$$\frac{d[A_1B]}{dt} = k_a[A_1][B] - k_d[A_1B] = 0 \tag{5.7}$$

$$\frac{d[A_2B]}{dt} = k_a[A_2][B] - k_d[A_2B] = 0 \tag{5.8}$$

$$\frac{d[A_3B]}{dt} = k_a[A_3][B] - k_d[A_3B] = 0 \tag{5.9}$$

$$\frac{d[A_4B]}{dt} = k_a[A_4][B] - k_d[A_4B] = 0 \tag{5.10}$$

As all of the reactions occur simultaneously, we can assume the concentration/ surface of available AuNPs [B] are the same for the Eqs. 5.7–5.10. Thus the above equations can be expressed as,

$$[A_1B] = \frac{k_{a1}[A_1][B]}{k_{d1}} = K_{a1}[A_1][B] \tag{5.11}$$

$$[A_2B] = \frac{k_{a2}[A_2][B]}{k_{d2}} = K_{a2}[A_2][B] \tag{5.12}$$

$$[A_3B] = \frac{k_{a3}[A_3][B]}{k_{d3}} = K_{a3}[A_3][B] \tag{5.13}$$

$$[A_4B] = \frac{k_{a4}[A_4][B]}{k_{d4}} = K_{a4}[A_4][B] \tag{5.14}$$

where, $K_a = k_a/k_d$. Thus the ratio of each protein adsorbed on the surface AuNPs can be expressed as,

$$[A_1B] : [A_2B] : [A_3B] : [A_4B] = K_{a1}[A_1] : K_{a2}[A_2] : K_{a3}[A_3] : K_{a4}[A_4] \tag{5.15}$$

5.4.5 *Studying Interactions of Serum Components with PEGylated AuNPs*

17 nm AuNPs were captured on the SPR chip surface (as described in Sect. 5.4.2) and stabilized for 30 min. Thiolated PEG-2000 (Lysanbio, AL, USA) was dissolved in water to a concentration of 100 mM and was injected onto the AuNPs modified SPR sensor surface. The injection time was 2 min followed by a stabilization step

of 10 min. For protein injection, all four chosen serum components HSA, fibrinogen, IgG and ApoA1 were dissolved in PBS, as in the previous section. Then the sensor chip was rotated by 90° and the four proteins were injected on to the sensor surface at a flow rate of 100 μL/min.

5.4.6 Studying Effect of PEG Chain Length on Corona Formation

17 nm AuNPs were captured on the SPR chip surface (as described above) and stabilized for 30 min. Thiolated PEG-2000, PEG-5000, PEG-10000, PEG-20000 and PEG30000 (Lysanbio, AL, USA) were dissolved in water, each separately, to a concentration of 100 mM each and were injected onto the AuNPs modified SPR sensor surface. The injection time was 2 min followed by a stabilization step of 10 min. 20% diluted human serum was injected after the chip was rotated through 90°. The interaction between serum and PEGylated AuNPs was made to occur for 2 min after which the injection of serum was stopped and lanes were flushed with PBST to observe the dissociation phase.

5.4.7 Studying the Interaction of Serum and Streptavidin with Model Hybrid NP Systems

In this experiment, we first make the following solutions: (a) 200 μM of SH-PEG-Biotin (M.W.5000), denoted as biotin in Fig. 5.5, (b) 200 μM of SH-PEG-Biotin (M.W.5000) and 100 μM SH-PEG (M.W.2000), denoted as Hybrid in Fig. 5.5, (c) 100 μM of SH-PEG (M.W.2000), denoted as PEG in Fig. 2.6.

Using previously mentioned methods, 17 nm AuNPs were captured on the SPR chip surface and stabilized for 30 min. A hybrid layer of surface functionalization material is formed on the AuNPs by co-injection of 100 μM of thiolated PEG-2000 (Lysanbio, US) and 200 μM of thiolated PEG-biotin (Thermos Scientific), both dissolved in water. Using the same protocol, two other lanes were modified using 100 μM of thiolated PEG-2000 and 200 μM of thiolated PEG-biotin, respectively. Human serum (Sigma-Aldrich, MO, USA) was diluted to 20% using PBS and injected onto the SPR Sensor. Streptavidin (Sigma-Aldrich, MO, USA) was dissolved in PBS buffer and injected onto the SPR sensor at a concentration of 100 μM. The injection times for both analytes were 2 min at a flow rate of 100 μl/min and sensor responses were recorded at 100 s after injection period. The sensor's inter-spots, which are the native surface of the sensor, were used as reference and all signals were subtracted from the reference signal.

References

1. Brewer, S.H., et al.: Probing BSA binding to citrate-coated gold nanoparticles and surfaces. Langmuir **21**, 9303–9307 (2005)
2. Brown, D.M., et al.: Interaction between nanoparticles and cytokine proteins: impact on protein and particle functionality. Nanotechnology **21**, 215104 (2010)
3. Bajaj, A., et al.: Stability, toxicity and differential cellular uptake of protein passivated-Fe3O4 nanoparticles. J. Mater. Chem. **19**, 6328–6331 (2009)
4. Boraschi, D., Costantino, L., Italiani, P.: Interaction of nanoparticles with immunocompetent cells: nanosafety considerations. Nanomedicine **7**, 121–131 (2011)
5. Cedervall, T.: Understanding the nanoparticle-protein corona using methods to quantify exchange rates and affinities of proteins for nanoparticles. Proc. Natl. Acad. Sci. U.S.A. **104**, 2050–2055 (2007)
6. Fadeel, B., Garcia-Bennett, A.E.: Better safe than sorry: understanding the toxicological properties of inorganic nanoparticles manufactured for biomedical applications. Adv. Drug Deliv. Rev. **62**, 362–374 (2010)
7. Casals, E., et al.: Time evolution of the nanoparticle protein corona. ACS Nano **4**, 3623–3632 (2010)
8. Dell'Orco, D., et al.: Modeling the time evolution of the nanoparticle-protein corona in a body fluid. PLoS ONE **5**, e10949 (2010)
9. Gasser, M.: The adsorption of biomolecules to multi-walled carbon nanotubes is influenced by both pulmonary surfactant lipids and surface chemistry. J. Nanobiotechnol. **8**, 31 (2010)
10. Cifuentes-Rius, A., et al.: Optimizing the properties of the protein corona surrounding nanoparticles for tuning payload release. ACS Nano **7**(11), 10066–10074 (2013)
11. Deng, Z.J.: Differential plasma protein binding to metal oxide nanoparticles. Nanotechnology **20**, 455101 (2009)
12. Kah, J.C.Y., et al.: Protein coronas on gold nanorods passivated with amphiphilic ligands affect cytotoxicity and cellular response to penicillin/streptomycin. ACS Nano **8**(5), 4608–4620 (2014)
13. Georgieva, J.V.: Surface characteristics of nanoparticles determine their intracellular fate in and processing by human blood-brain barrier endothelial cells in vitro. Mol. Ther. **19**, 318–325 (2011)
14. Ghosh, P.: Intracellular delivery of a membrane-impermeable enzyme in active form using functionalized gold nanoparticles. J. Am. Chem. Soc. **132**, 2642–2645 (2010)
15. Iversen, T.G., Skotland, T., Sandvig, K.: Endocytosis and intracellular transport of nanoparticles: present knowledge and need for future studies. Nano Today **6**, 176–185 (2011)
16. Docter, D., et al.: Quantitative profiling of the protein coronas that form around nanoparticles. Nat. Protocols **9**(9), 2030–2044 (2014)
17. Mahmoudi, M., et al.: Temperature: the "ignored" factor at the nanobio interface. ACS Nano **7**(8), 6555–6562 (2013)
18. Jedlovszky-Hajdú, A., et al.: Surface coatings shape the protein corona of SPIONs with relevance to their application in vivo. Langmuir **28**(42), 14983–14991 (2012)
19. Lesniak, A., et al.: Effects of the presence or absence of a protein corona on silica nanoparticle uptake and impact on cells. ACS Nano **6**(7), 5845–5857 (2012)
20. Tenzer, S., et al.: nanoparticle size is a critical physicochemical determinant of the human blood plasma corona: a comprehensive quantitative proteomic analysis. ACS Nano **5**(9), 7155–7167 (2011)
21. Monopoli, M.P., et al.: Physical—chemical aspects of protein corona: relevance to in vitro and in vivo biological impacts of nanoparticles. J. Am. Chem. Soc. **133**(8), 2525–2534 (2011)
22. Walczyk, D., et al.: What the cell "sees" in bionanoscience. J. Am. Chem. Soc. **132**(16), 5761–5768 (2010)

23. Cedervall, T., et al.: Understanding the nanoparticle–protein corona using methods to quantify exchange rates and affinities of proteins for nanoparticles. Proc. Natl. Acad. Sci. **104**(7), 2050–2055 (2007)
24. Liedl, T., et al.: Fluorescent nanocrystals as colloidal probes in complex fluids measured by fluorescence correlation spectroscopy. Small **1**(10), 997–1003 (2005)
25. Nienhaus, G.U., Maffre, P., Nienhaus, K.: Chapter four—studying the protein corona on nanoparticles by fcs, in methods in enzymology. In: Sergey, Y.T. (ed.) pp. 115–137. Academic Press, Cambridge (2013)
26. Zhang, F., et al.: Polymer-coated nanoparticles: a universal tool for biolabelling experiments. Small **7**(22), 3113–3127 (2011)
27. Cho, E.J., et al.: Nanoparticle characterization: state of the art, challenges, and emerging technologies. Mol. Pharm. **10**(6), 2093–2110 (2013)
28. Canovi, M., et al.: Applications of surface plasmon resonance (SPR) for the characterization of nanoparticles developed for biomedical purposes. Sensors (Basel, Switzerland) **12**(12), 16420–16432 (2012)
29. Aggarwal, P., et al.: Nanoparticle interaction with plasma proteins as it relates to particle biodistribution, biocompatibility and therapeutic efficacy. Adv. Drug Deliv. Rev. **61**(6), 428–437 (2009)
30. Karmali, P.P., Simberg, D.: Interactions of nanoparticles with plasma proteins: implication on clearance and toxicity of drug delivery systems. Expert Opin. Drug Deliv. **8**(3), 343–357 (2011)
31. Göppert, T., Müller, R.: Adsorption kinetics of plasma proteins on solid lipid nanoparticles for drug targeting. Int. J. Pharm. **302**(1), 172–186 (2005)
32. Tsai, D.-H., et al.: Adsorption and conformation of serum albumin protein on gold nanoparticles investigated using dimensional measurements and in situ spectroscopic methods. Langmuir **27**(6), 2464–2477 (2011)
33. Mahmoudi, M., et al.: Variation of protein corona composition of gold nanoparticles following plasmonic heating. Nano Lett. **14**(1), 6–12 (2013)
34. Núñez, S., Venhorst, J., Kruse, C.G.: Target–drug interactions: first principles and their application to drug discovery. Drug Discovery Today **17**(1), 10–22 (2012)
35. Urbinati, C., et al.: αvβ3-integrin-dependent activation of focal adhesion kinase mediates NF-κB activation and motogenic activity by HIV-1 Tat in endothelial cells. J. Cell Sci. **118** (17), 3949–3958 (2005)
36. Chu, R., Reczek, D., Brondyk, W.: Capture-stabilize approach for membrane protein SPR assays. Sci Reports **4** (2014)
37. Otsuka, H., Nagasaki, Y., Kataoka, K.: PEGylated nanoparticles for biological and pharmaceutical applications. Adv. Drug Deliv. Rev. **55**, 403–419 (2003)
38. Kim, H.R.: Analysis of plasma protein adsorption onto PEGylated nanoparticles by complementary methods: 2-DE, CE and protein lab-on-chip system. Electrophoresis **28**, 2252–2261 (2007)
39. Hatakeyama, H., Akita, H., Harashima, H.: The polyethyleneglycol dilemma: advantage and disadvantage of PEGylation of liposomes for systemic genes and nucleic acids delivery to tumors. Biol. Pharm. Bull. **36**(6), 892–899 (2013)
40. Jeon, S.I., et al.: Protein—surface interactions in the presence of polyethylene oxide: I. Simplified theory. J. Colloid Interface Sci. **142**(1), 149–158 (1991)
41. Kah, J.C.Y., et al.: Critical parameters in the pegylation of gold nanoshells for biomedical applications: an in vitro macrophage study. J. Drug Target. **17**(3), 181–193 (2009)
42. McPherson, T., et al.: Prevention of protein adsorption by tethered poly(ethylene oxide) layers: experiments and single-chain mean-field analysis. Langmuir **14**(1), 176–186 (1998)
43. Gref, R., et al.: 'Stealth' corona-core nanoparticles surface modified by polyethylene glycol (PEG): influences of the corona (PEG chain length and surface density) and of the core composition on phagocytic uptake and plasma protein adsorption. Colloids Surf., B **18**(3–4), 301–313 (2000)

44. Pozzi, D., et al.: Effect of polyethyleneglycol (PEG) chain length on the bio-nano-interactions between PEGylated lipid nanoparticles and biological fluids: from nanostructure to uptake in cancer cells. Nanoscale **6**(5), 2782–2792 (2014)
45. Frens, G.: Controlled nucleation for the regulation of the particle size in monodisperse gold suspensions. Nature **241**, 20–22 (1972)
46. Nahshol, O., et al.: Parallel kinetic analysis and affinity determination of hundreds of monoclonal antibodies using the ProteOn XPR36. Anal. Biochem. **383**(1), 52–60 (2008)

Chapter 6
Conclusion

Abstract This chapter provides the concluding remarks on the work we have done along with an outlook on the future work possible in this domain.

No work in science is beyond improvement. On our part, we believe we have demonstrated techniques to make measurements possible where they were none previously and accurate where they were inaccurate and error-prone. But, just as we have stated earlier, there is always more work to be done and some improvement to be made.

The technique for ascertaining the hydrophilicity/hydrophobicity of a solid surface has been the contact angle of water method for a very long time. Through advancements in precision syringes, motors with very precise pitch and optics, the technique has, of course become more accurate over the years. However, with sciences being more intent towards traversing down the length scales from macro to micro to nano, this well-proven technique has been faced with a seemingly unsurmountable problem. Unless we generate extremely small droplets in the nano or picolitre regime, the resultant data acquired will have inherent artefacts resulting from the averaging of various chemical heterogeneities on the surface. Herein, lies our breakthrough. The use of force-distance curves to quantify hydrophilicity of surfaces is a simple technique which can be very useful for obtaining local surface energies from extremely small areas on surfaces. Since the tip of an AFM cantilever is usually around 5–10 nm in diameter, we would in fact be able to probe such small areas without the need for expensive equipment which generate nano/picolitre droplets for extremely complicated nano/picolitre level contact angle measurements. The technique of deriving hydrophilicity from the interaction of the AFM tip from the native water layer on the surface is also affected by the material used to fabricate the tip. Hence, the logical next step is to acquire F-D curves over the same samples with different tips made of standard materials used by the AFM probe manufacturers. When this data is available, it will become possible to generate universal curves for each tip which can be referred to readily to find the probable local surface energy and contact angle for a particular unknown sample being explored with a known tip.

A. Patra, *Quantifying Interactions of Biomolecules with Inorganic Surfaces*,
Springer Theses, DOI 10.1007/978-3-319-30728-2_6

The impact of the study of interactions between phospholipids and metal oxides is perhaps the most immediate. In general, through our own endeavors, we have found that it is difficult to control all the parameters which influence interaction of metal oxides and lipid species. Reports in literature do talk about successes in selectively enriching non-phospholipid fractions of lipidomes by trapping phospholipids. However, no attempt has been made to understand and control the various parameters which exert their own influences. Primary among them is the actual contact area of metal oxides available for phospholipid capture. There is a very interesting conundrum here. Ideally, one would need to have smallest possible particles to maximize the available surface area. However, as the particles become smaller, they pack closer and closer thereby generating very high backpressures in the column. High backpressures also slow down liquid flow through the column which changes the amount of time phospholipids spend in contact with the resins. This contributes to experimental error. Through rigorous experimentation, we have found the lower limit of particle size, beyond which the fabrication of a working μcolumn is impossible. The pressure required to push liquids through such μcolumn is enough to completely rupture the μcolumn as such. The surface plasmon resonance based technique for studying phospholipid adhesion to oxide surfaces has been a novel approach to the field of solid phase extraction resin design. By bringing contact area and flow rate under experimental control, we have been able to precisely quantify the amount of phospholipids that adhere to the oxide surfaces meant to act as solid phase extraction resins. Another off-shoot of this approach is the easy way to know if a particular oxide etches away under the conditions used to bind phospholipids. Under the binding conditions generally used, we found 5 oxides that form salts which are detrimental to the health of mass spectrometry systems apart from causing suppression of ionization of lipid species. From our data, we have outlined 3 oxides which can be used for the purpose of phospholipid capture. All of them are from the 3d period. It would be interesting to see if the same trend is maintained in the 4d or 5d period. Since rare earth oxides are also d^0 systems, efforts can also be directed towards studying their interaction with phospholipids. However, unless a significant increase in performance is obtained, it might not be financially profitable to use 4d oxides or rare earth oxides since they are generally more expensive than the best performing 3d counterparts. Knowledge of phospholipid-metal oxide interactions will ultimately help in the fabrication of fraction enrichment devices in a variety of formats which are completely compatible with the workflow which has been around for more than two decades.

The multiplexed surface plasmon resonance based technique for studying serum proteins interacting with nanoparticles is an exciting addition to the tools being widely used today. Though nano-bio interfaces have been a hot area of scientific investigation over the last 7–8 years, the existing tools have limitations. Simple techniques like DCS, FCS, DLS and Zeta potential evaluation are fast in terms of their turnaround time. However the data varies highly across experimental runs and hence show high error bars. More comprehensive techniques like SDS-PAGE/MS are accurate but are very labor-intensive and the large number of steps involved in collecting meaningful data introduces the possibility of user-dependent artefacts.

Besides, none of the existing techniques are able to furnish true kinetic parameters of serum interactions with engineered nanomaterials in real time. By identifying all these deficiencies of all techniques and imagining what a reasonable solution to fill in the gap would be, we have been able to establish a robust protocol which, we are sure, will be used by researchers in the field immediately. With our technique, we have been able to obtain kinetic parameters of interactions of proteins with engineered nanosystems. These are of critical importance to researchers involved in the design of nano drug delivery systems (NDDS). The modular nature of the protocol enables users to choose any engineered nanomaterial and a corresponding chemical linker to capture it on the sensing surface. Further work would be directed towards studying interactions of more clinically relevant nanomaterials, such as SiO_2, Fe_3O_4, carbon nanotubes, PolyLactic Acid, liposomes and others. Moreover, using this technique, one can test the efficiency of anti-biofouling measures employed to prevent alteration of identity of the NDDS while still being functional enough to identify its intended target and deliver the packed cargo. This translates into significant savings on time and money, when only the most promising candidates from this initial screening, which takes only few hours to finish, are taken forward to the next round of rigorous testing.

In conclusion, it isn't an overstatement to say, we have indeed been able to quantify interactions of biomolecules with inorganic surfaces. We would like to close this work with this beautiful quote from one of the most creative minds of the 20th century which captures the spirit with which this work has been accomplished.

> Around here, however, we don't look backwards for very long. We keep moving forward, opening up new doors, and doing new things, because we're curious and curiosity keeps leading us down new paths.
>
> Walter Elias "Walt" Disney.
>
> Disneyland Paris—The Family Guide (2007), opening page.

Appendix A
SPR Sensorgrams for Interactions of 3 Serum Proteins with Gold Nanoparticles

Further from Sect. 5.3.2 (Figs. A.1, A.2 and A.3).

A. Patra, *Quantifying Interactions of Biomolecules with Inorganic Surfaces*,
Springer Theses, DOI 10.1007/978-3-319-30728-2

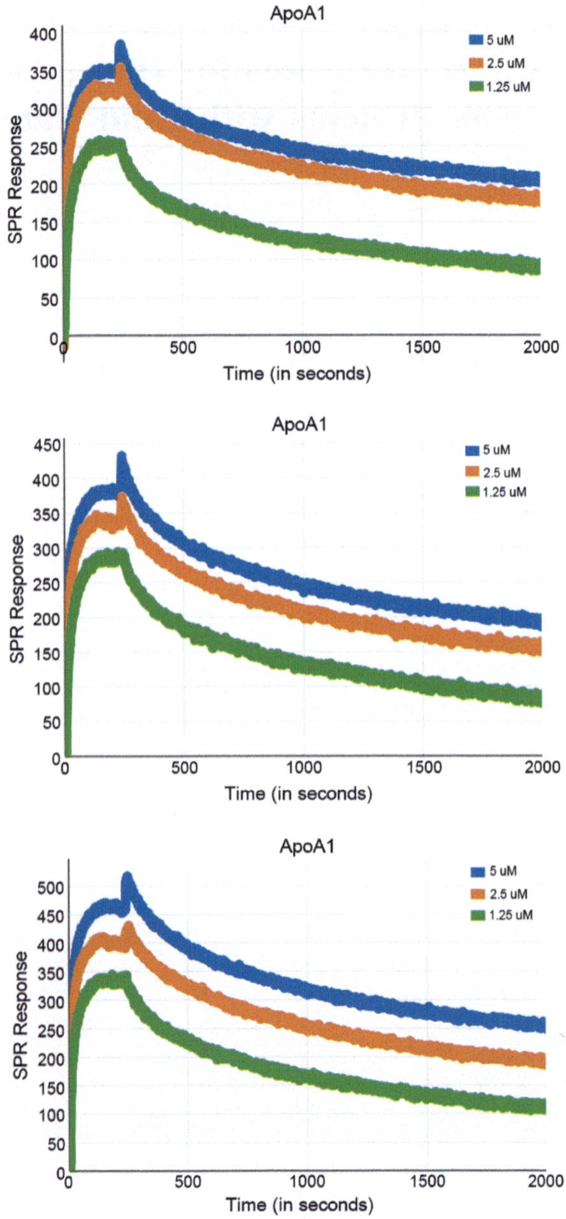

Fig. A.1 SPR Sensorgram for ApoA1 interacting with AuNPs. Different colours represent different concentrations. Concentrations shown in legend. 3 sets of data have been shown

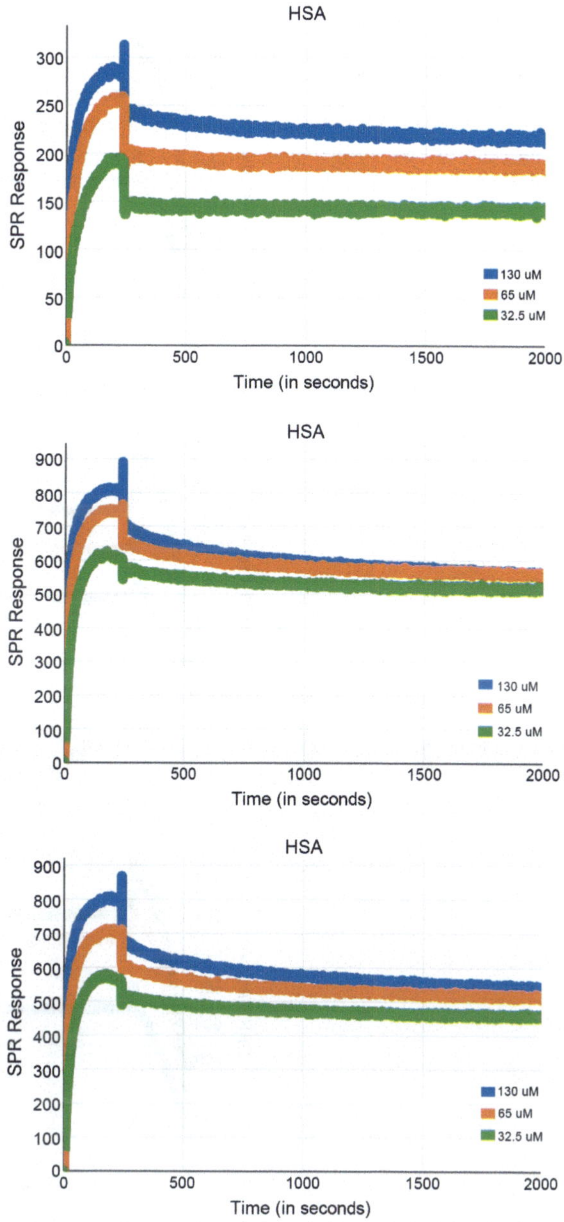

Fig. A.2 SPR Sensorgram for HSA interacting with AuNPs. Different colours represent different concentrations. Concentrations shown in legend. 3 sets of data have been shown

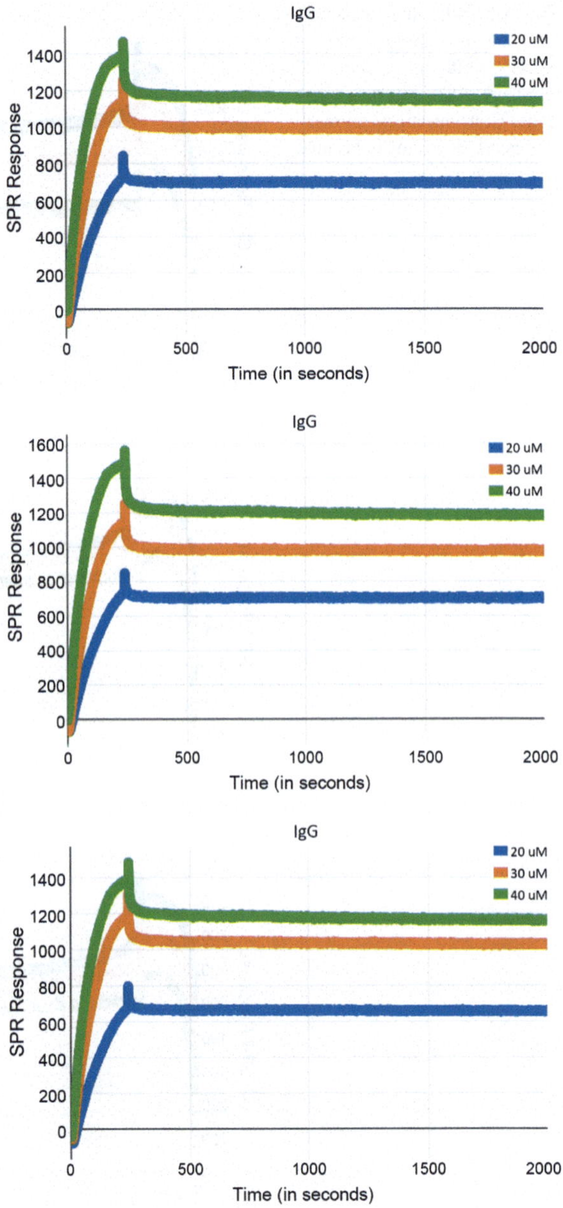

Fig. A.3 SPR Sensorgram for IgG interacting with AuNPs. Different colours represent different concentrations. Concentrations shown in legend. 3 sets of data have been shown

Appendix B
SPR Sensorgrams for Interactions of PEG with Chemically Active Chip Surface

See Fig. A.4.

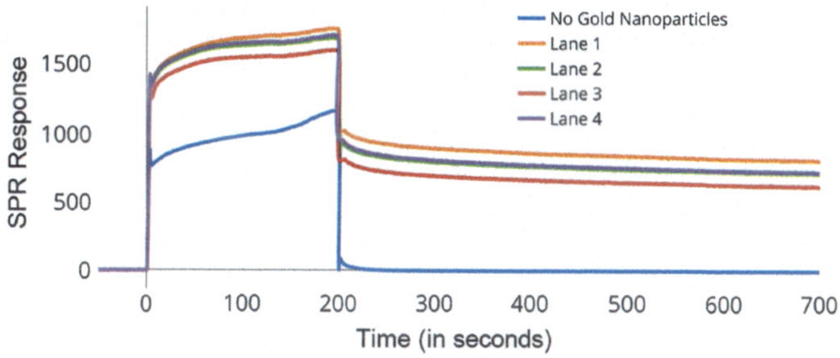

Fig. A.4 Control studies to show no interaction of PEG with the chemically active linker surface. Lanes 1, 2, 3 and 4 have gold nanoparticles and PEG is seen to adsorb onto the captured nanoparticles. When there are no captured gold nanoparticles, PEG doesn't adsorb at all onto the chip surface

A. Patra, *Quantifying Interactions of Biomolecules with Inorganic Surfaces*,
Springer Theses, DOI 10.1007/978-3-319-30728-2

Curriculum Vitae

Dr. Abhijeet Patra

NUS Nanoscience and Nanotechnology Institute
#11-01, T-Lab
5A Engineering Drive 1
Singapore 117411
Email: abhijeetpatra88@gmail.com

Professional Profile
A researcher working in the area of bio-nanotechnology with expertise in studying interactions of biomolecules with organic/inorganic surfaces or assemblies and tailoring cell/bacterial growth by using metal oxide thin films.

Education and Qualifications

Ph.D.	NUS Graduate School for Integrative Sciences and Engineering, National University of Singapore (Aug 2011–Nov 2015)
Bachelor of Technology	Electrical and Electronics Engineering, National Institute of Technology, Tiruchirappalli, India (2007–2011)

Graduate Projects:

- Development of new protocols/technologies to study interaction of biomolecules with organic/inorganic surfaces or assemblies for the purposes of therapeutics delivery and purification of selected groups of molecules.
- Development of new protocols/technologies for rapidly screening materials for cell/bacterial growth.

Publications:

1. Patra, Abhijeet, Tao Ding, Gokce Engudar, Yi Wang, Michal Marcin Dykas, Bo Liedberg, James Chen Yong Kah, Thirumalai Venkatesan, and Chester Lee Drum. "Component-Specific Analysis of Plasma Protein Corona Formation on Gold Nanoparticles Using Multiplexed Surface Plasmon Resonance." *Small* (2015).

A. Patra, *Quantifying Interactions of Biomolecules with Inorganic Surfaces*,
Springer Theses, DOI 10.1007/978-3-319-30728-2

2. Sarkar, T.P., K. Gopinadhan, M. Motapothula, S. Saha, Z. Huang, S. Dhar, A. Patra et al. "Unexpected observation of spatially separated Kondo scattering and ferromagnetism in Ta alloyed anatase TiO_2 thin films." *Scientific reports* 5 (2015).
3. Li, J., Yoong, S.L., Goh, W.J., Czarny, B., Yang, Z., Poddar, K., Dykas, M.M., Patra, A., Venkatesan, T., Panczyk, T. and Lee, C., 2015. In vitro controlled release of cisplatin from gold-carbon nanobottles via cleavable linkages. *International journal of nanomedicine*, *10*, p. 7425.

Positions of responsibility

Oct 2014	Coordinator, NanoSpark

Hosting, managing and coordinating events for students, faculty members and entrepreneurs. Events include talks, interactive sessions, networking dinners.

Jan 2014–May 2014	Teaching assistant, Bioengineering, NUS

Handled tutorials and laboratory sessions for an undergraduate module on BioSignal Processing, taught by Asst. Prof. James Kah.

Aug 2010–March 2011	Head, Events, Pragyan

Selecting and managing events to be conducted at Pragyan, a national level technical festival in India with a budget of S$210,000.

Awards

- NGS scholarship to pursue graduate studies at National University of Singapore, Aug 2011 to Aug 2015
- NUS Enterprise Practicuum Grant 7.5K SGD for the development of a prototype
- Best Oral Presentation, Life Opportunities Research Symposium 2014, NUS
- Runners up at ASEAN Science Slam 2013 organised by EURAXESS, 2013
- Indian National Academy of Engineers Fellowship, May to July 2010.

The manufacturer's authorised representative in the EU is Springer Nature Customer Service Centre GmbH, Europaplatz 3, 69115 Heidelberg, Germany. If you have any concerns regarding our products, please contact ProductSafety@springernature.com

Printed and bound by CPI Group (UK) Ltd, Croydon, CR0 4YY
12/08/2026
02194544-0001

24/7
Arbeit zwischen Sinnstiftung
und Entgrenzung
Work between meaning
and imbalance

Vorwort

Andreja Hribernik

Die Ausstellung *24/7* bietet eine Reihe von Perspektiven auf die Arbeit und auf die Einbindung der*des Einzelnen in Prozesse, die von der Arbeit definiert werden. Wie der Untertitel der Ausstellung andeutet, schwankt die gesellschaftliche und individuelle Einstellung zur Arbeit oft zwischen zwei Extremen, zwischen Sinnstiftung und Entgrenzung. Alle Werke widerspiegeln trotz ihrer unterschiedlichen Sichtweisen auf das Thema eine gewisse Unvermeidbarkeit dieser Beziehung zur Arbeit. Die eigentliche Negation und die Perspektive der Faulheit werden durch die Abwesenheit von Arbeit definiert. Diese Ambivalenz und die Unmöglichkeit, davon abzurücken, ist eng mit dem gesellschaftlichen System verbunden, das alle unsere Beziehungen und Prozesse kapitalisiert.

Die Zeit spielt auch bei der Arbeit eine wichtige Rolle, durch die der Wert unserer Arbeit, unsere Effizienz und unser Anspruch auf Bezahlung gemessen werden, und die Kontrolle der Zeit sowie die Zählung der Zeit als System haben die sozialen Beziehungen vollständig durchdrungen. Zuweilen bietet die Ausstellung eine fast dystopische Vision einer totalen sozialen Kontrolle, die nicht durch ein politisches, sondern durch ein wirtschaftliches System realisiert wurde. Diese Vision führt uns zu der Tatsache, dass wir durch die Ausstellung sehr leicht die Strukturen verstehen können, innerhalb derer wir uns im Alltag bewegen.

Mitunter begegnet uns in der Ausstellung eine harsche Kritik am bestehenden System, gerade durch die Perspektive feministischer Positionen, die auf die immer noch anhaltende Sinnhaftigkeit des Widerstands hinweisen und die Möglichkeiten eines Auswegs aufzeigen. Vielleicht kann der Kampf der Frauen um bezahlte Hausarbeit zu einer zentralen Forderung werden, die das bestehende Gesellschaftssystem grundlegend erschüttern könnte? Wie Silvia Federici 1974 in ihrem Text *Lohn gegen Hausarbeit* vorschlug, besteht die Lösung für die Stellung der Frau in der Gesellschaft nicht in einem Kampf für höhere Löhne und ein unterstützendes Sozialsystem mit Kinderpflege, Wäschereien und organisierten Mahlzeiten, sondern in der Forderung nach bezahlter Hausarbeit. Wie Federici reflektiert, ist die Forderung nach bezahlter Hausarbeit ein Kampf, um den Plan zu durchbrechen, den das Kapital für die Frauen aufgestellt hat und der gleichzeitig eine Spaltung innerhalb der Arbeiterklasse bewirkt, die es dem Kapital ermöglicht, seine Macht zu erhalten.

Das Thema Frauenarbeit wird auch in der Ausstellung *Alles Arbeit* in der Partnerinstitution Museum für Geschichte aufgegriffen, die sich anhand von fotografischen Bildern mit der Wahrnehmung von Frauen in der Gesellschaft auseinandersetzt und trotz der Fokussierung auf Archivmaterial aus den 1950er- und 1960er-Jahren relevante Fragen der Gegenwart aufwirft.

Die Ausstellung *24/7. Arbeit zwischen Sinnstiftung und Entgrenzung* ist die zentrale Ausstellung im Jahr 2024, mit der wir langfristige Programmthemen im Kunsthaus setzen, und sie ist auch inhaltlich mit dem übrigen Programm verknüpft und zum Teil als Prozess konzipiert, in dem einige künstlerischen Arbeiten noch im Laufe der Ausstellungsdauer realisiert und abgeschlossen werden.

→ Stimmen aus der Arbeitswelt

Zu *24/7* hat das Vermittlungsteam unterschiedliche Personen eingeladen, über die Welt der Arbeit zu reflektieren. Ausgehend von gezeigten Arbeiten ergänzen aufgezeichnete Gespräche die Ausstellung um ein breites Spektrum von Erfahrungen in der und rund um die Arbeitswelt.

Theresa Hattinger, Michael Hieslmair, Michael Zinganel
→ S. 28 / p. 29

Jeremy Deller
→ S. 12 / p. 13

Foreword

Andreja Hribernik

The exhibition *24/7* offers a range of views on work and how the individual engages with the processes defined by work. As suggested by the show's subtitle, the attitudes of society and the individual towards work often fluctuate between two extremes — between meaning and imbalance. Although they offer different perspectives on the subject, all of the artworks reflect a certain inevitability to this relationship with work. The actual negation and the aspect of laziness are defined by the absence of work. This ambivalence, together with the impossibility of distancing ourselves from it, is deeply embedded in a social system that capitalises all of our relationships and processes.

Time also plays an important role in work: the value of our work, our efficiency and our entitlement to remuneration are measured in time, and the control of time and the counting of time as a system have entirely permeated social relations. At times, this exhibition offers an almost dystopian vision of total social control that is realised through an economic rather than a political system. This vision brings us to the fact that the structures within which we move during everyday life are clearly revealed by the exhibition.

At times, we encounter in the show a harsh critique of the existing system. This occurs especially through the lens of feminist positions, which point to the enduring meaningfulness of resistance and offer possible ways out. Perhaps the women's campaign for domestic labour to be paid can become a key demand that would fundamentally shake up the existing social system? Silvia Federici proposes in her 1974 text *Wages Against Housework* that the solution to women's status in society is not to call for higher wages and a supportive welfare system with childcare, laundries and organised meals, but instead to demand that housework should be paid. As Federici reflects, the push for paid domestic work is a fight to shatter the plan that capital has made for women, a plan which also creates divisions within the working class that allow capital to hold on to its power.

The theme of women's work is also explored in *It's All Work* at our partner institution, the History Museum. This exhibition uses photographic images to examine the perception of women in society. Although the show's focus is on archive materials from the 1950s and 1960s, it also raises pertinent questions about the present day.

24/7. Work between meaning and imbalance is the central exhibition running in 2024 and forms the basis for long-term programme themes at the Kunsthaus. Its content also ties in with the rest of our programme. A number of artworks will be implemented and completed during the course of the show, which is partly conceived as a process.

I would like to thank the curator Katia Huemer and assistant curator Martin Grabner as well as the entire Kunsthaus team, together with other departments at the Universalmuseum Joanneum who have played a role in creating this exhibition.

Peter Kogler →→
→ S. 34 p. 35

Antje Ehmann & Harun Farocki
→ S. 14 / p. 15, S. 18 / p. 19

No, I didn't send it.
Listen, I'm getting ready for the

Alles Gute zum Alltag

Agnes Maier

Es beginnt mit einem Röcheln.
Ein kleines Kratzen im Hals,
ein Nieser, der durch Bürogänge schallt,
es drehen sich Köpfe, manch ein Mund spricht:
Gesundheit!
Und ob's eh nicht was Ansteckendes ist?

Das Jahr 2020 gab uns vieles an die Hand:
Dass man zum Überleben hauptsächlich Nudeln und Klopapier braucht und
ein ganz neues Verständnis von Krankenstand:

Dass man daheimbleiben sollte, wenn man krank ist!

Das ist ja oft immer noch nicht so richtig klar:

Ohne behördlich ausgestelltes Daheimbleibedokument in Form eines Absonderungsbescheids heißt „krank sein“ nicht automatisch auch „arbeitsbefreit“.

Ein grippaler Infekt, aber der Antigen-Test war negativ? — Aso, na dann!

Ein Schleudertrauma? —
Ich mein, da besteht ja noch nicht mal Ansteckungsgefahr!
Und du —
Du wirkst so ausgebrannt, es hängt Asche in deinem Haar!
Aber bitte, das ist ja kein Grund zum Daheimbleiben!
Das ist doch super, wenn du gefragt bist!

Weil das halt ein Garant für Kapital ist.

Dreifach beringte Augenhöhlen, eingefallen und aschfahl,
zeugen nur von harter Arbeit
und deiner Willenskraft aus Stahl,
sogar noch dein Bedürfnis nach Schlaf zu überwinden.
Das finden Leute „fleißig“!
Weißt du's noch nicht? Drogensucht war gestern! Wir haben jetzt was Besseres!
Um den eigenen Körper mit Nachdruck zugrunde zu richten, ist es jetzt nicht mehr nötig, mit Substanzen zu dealen,
das geht schon mit „Karriere“!
Ja und 'nem Haufen Koffein.

Es scheint, als generiere man Begeisterung
nur durch Leistung,
die Bewunderung der Menschen
nur durch konstante Überschreitung der eigenen Grenzen.
Unmenschliches kriegt den meisten Applaus,
wenn du leidest und dich quälst, bleibt der Beifall nicht aus.
Dieses Land ist so katholisch, es ist mir ein Graus!

Und was soll das eigentlich heißen, „Vollzeit erwerbstätig“?
Jetzt mal ehrlich ...

Der Mensch, der heute noch die 40-Stunden-Woche voll supergut machbar findet, ist doch garantiert entweder Single mit latenter Karrierefixierung ODER ein weißer Mann mittleren Alters, der seiner gebärfähigen Vollzeit-Raumpflegemanagerin mit Zusatzkompetenz Kinderbetreuung mit 20 einen Ring an den Finger gesteckt hat, um dann natürlich zu finden:
Och, wenn's weiter nichts ist, sind die 5 x die Woche 8—12 Stunden Arbeit am Tag ja echt kein großes Ding mehr. Und am Wochenende spiel ich eh gern mal ein bisschen Fußball im Garten.
Oder wie soll das sonst möglich sein?

Jetzt mal ehrlich!

Welcher Mensch, der 40 Stunden die Woche arbeitet, hat:
einen gepflegten Wohnraum, Kinder, die er besser als nur vom Gute-Nacht-Sagen kennt, einen gefüllten Kühlschrank, hier und da ein bisschen Zeit für sich selbst *und* keine Hilfe daheim?
Das kann sich nicht ausgehen!

Aber „Was der leistet!", „Was die schafft!" —
Heißts meist leider nur, wenn das „Was" auch Kapital erbracht hat oder sonst wie beigetragen hat zur Schieflage der Welt,
in der nur Macht und Geld und Einfluss unter „was erreicht haben" fällt.

Aber in einem Affenzirkus, in dem
„das Hamsterrad" die Hauptattraktion darstellt, will ich euch mal was zeigen:
zwei Schritte zurückmachen und einfach mal —

Schweigen.

Erzeugt da draußen schon so viel mehr Aufmerksamkeit,
als jeder wild gewordene Affe, der aus voller Kehle schreit.

Ich weiß nicht wie's euch geht, aber aus meiner Position
ist in einer so lauten Welt Stille eine Art Revolution.
Und langsame Bewegungen wie Sabotage betreiben,

ich will von der Karriereleiter runterspringen, um am Boden der Tatsachen zu bleiben
und dort ein bisschen Konfetti zu verstreuen
für banale Dinge, die gar nicht so einfach sind,
wie alle immer meinen,
oder nur weil sie ALLE machen nicht mehr wichtig genug scheinen.

Aber ich – Ich zum Beispiel bin ein Morgenmuffel und muss manchmal vor 7 Uhr raus.
Und da wär's schon manchmal nett, bekäme ich ein bisschen Applaus.
Einfach nur fürs Mühe-Geben!

Herzliche Gratulation zum heutigen Erwachen, und zwar nicht das geistige, sondern einfach das in der Früh, für das du diesmal nur 3 Schlummermoduszyklen statt der üblichen 5 bis 7 gebraucht hast.
Gut gemacht. 1 A!

Danke! Ja, danke.

Ein bisschen Wertschätzung für den Alltag
ist Beruhigung für den Herzschlag.

Und wie geht es dir?
Und zwar wirklich.
Nicht nur als Floskel, zur Erfüllung der Gesellschaftsetikette
ist das keine nette, beiläufige Frage mehr.
Nein, die ist schwer.

In welchem Takt schlägt dein Herz?
Auf welcher Frequenz spielt dein Schmerz?
Und ist die Zufriedenheit noch lauter als er?

Kommst du auf ein solides Wohlbefinden?

Kurz: Wo ist deine Happiness-Baseline zu finden?

Denn es muss nicht jeder Tag der geilste deines Lebens sein.
(Mal ehrlich – wie anstrengend wär so viel Adrenalin?)
Aber jeder kleine Stein im Schuh drückt einfach weniger schlimm,
wenn die Füße nicht schon vorher voller Blasen waren.
Das ist eine Tatsache.

Wen interessiert's, wie viel du leistest oder Profit du generierst,
wenn's dir scheiße dabei geht und du die Freude verlierst
und du zu wenig schläfst und dir der Stress alles andre nimmt –

Wir müssen aufhören, Menschen zu beklatschen,
weil sie scheiße zu sich selber sind!

Denn das ist scheiße und echt für niemanden fair.
Wenn man ertrinkt in Cortisol, fällt das Atmen leider schwer.

Wenn man feststeckt in Systemen, die auf Menschen vergessen,
weil sie immer nur versuchen, aus ihnen Wirtschaftsleistungen zu pressen.
Und am Ende sollen sie dann vielleicht noch froh sein,
wenn es ganz okay zum Leben reicht, während aus starren Pupillen Lebensfeuer entweicht,
langsam ausgeht, ausbrennt,
die Begeisterung verpufft, wo ich hinseh', ist Erschöpfung,
es liegt Asche in der Luft!

Und deshalb wünsch' ich euch von Herzen alles Gute zum Alltag
und will Konfetti dalassen
fürs Freundlich-zu-sich-selber-Sein
und Das-Beste-aus-manchen-Tagen-Machen.
Fürs Genug-Sein, wie man ist, ohne immer was zu leisten.
Fürs Füße-Hochlegen und Mal-auf-den-Kapitalismus-Scheißen.

Und für die Erkenntnis, dass es eben mehr als nur „Erwerbsarbeit" gibt
und dass der Schlüssel zur Gleichstellung nicht darin liegt – gar

nicht darin liegen kann! –,
jetzt einfach alle Menschen in ein männliches Rollenbild zu quetschen
und bis zur vollständigen Kiefersperre die Zähne zu fletschen.

Deshalb Cheers! auf den gesunden Egoismus und über diese
Verbissenheit lachen,
weil alle hier rennen, vergeb' ich „Thumbs-up!" fürs Langsam-Machen.

Und fürs Ehrlich-leise-zu-sich-selber-Sagen:

Ich bin es wert, auf mich zu achten und mich selber gern zu haben.

Liam Gillick
→ S. 22 / p. 23

Bitte nicht betreten
Please do not enter

Andreas Gursky
→ S. 26 / p. 27

Andreas Gursky

Lisa Großkopf
→ S. 24 / p. 25

Christoph Schwarz
→ S. 56 / p. 57

Happy day of the week

Agnes Maier

It starts with a rattling in your chest.
An itch in your throat
that sneeze that is heard all through the office
that turns heads, and somewhere someone says
Gesundheit!
Let's hope it's nothing infectious!

The year 2020 taught us many things:
Things like that all you need to survive are noodles and toilet paper,
as well as an entire new meaning of sick leave:

You should stay at home if you are sick!

Often this still isn't entirely clear:

If you don't have an official document that proves you have to stay at home

'being sick' does not automatically mean you are 'free from work'.

You feel like you are coming down with something, but the Antigen test was negative — well in that case!

Suffering from whiplash? —
It's not like that can be infectious!
And you —
You seem burned out, you have ash in you hair!
Come on, that can't be a reason for you to stay home!
It's great when you're in demand!

Because that is that guarantee for capital.

Eyes with rings like a tree, sunken and ashen
only go to show your hard work
your iron will
your ability to overcome your need for sleep.
That is what makes you industrious and hard working!
Have you not heard? — Drugs are a thing of the past! We have something bettcr now!
In order to break your body there is no more need to deal in substances,
you can now do all that with your 'career'!
And of course with lots of caffeine.

It almost seems as though you can only create excitement
through output,
spark the admiration of people
only through the constant overstepping of your own boundaries.
Inhumanity garners the most applause,
whenever you suffer, or torture yourself, that is when people will cheer.
This country is so Catholic, it gives me the fear!

And what is that even supposed to mean, to be 'full-time employed'
Let's be honest ...

The person who still finds the 40-hour week totally feasible today
is guaranteed to be either a single with a latent career fixation
OR a middle-aged white man who puts a ring on the finger of his
childbearing full-time room care manager with additional childcare

skills at the age of 20, only to observe:
Oh well, if it's nothing more, then working 8—12 hours a day for 5 days a week is really no big deal. And at the weekend I do like to play a bit of football in the garden anyway.
Or how else could that be possible?

Let's be honest!

Who is this person who works for 40 hours the week and still has:
A well kept living space, children who they know better than from just saying good night, a well-stocked fridge, and every now and again a little time for themselves *without* any help at home?
There is simply no way that can work!

But 'the things he can do', 'the things she has achieved'! —
That is all that is ever said, when 'what you do' brings in capital
or contributes in any other way to the skewed world we live in,
a world in which power and money and influence are the only
things that 'can be achieved'.

But in this circus in which,
the 'rat race' is the main attraction, I want to show you something:
take two steps back, and for once —

Quiet.

That in itself can draw so much more attention
than any scream at the top of your lungs could ever hope.

I don't want to speak for anyone else, but coming from my position
in this loud world, being quiet is a form of revolution.
To slow down becomes sabotage,
I want to jump off the career ladder, to stay grounded,
and there I will scatter glitter
for those simple things that aren't as simple,
as people always say,
or just because EVERYONE does them, they seem no longer
important enough.

But me — I, for example, am absolutely not a morning person, but have to get up at 7 a.m. sometimes.
And for that, it would be nice to be get some applause.
Just for doing my best!

Congratulations for your awakening, not the spiritual kind, but simply the kind you do in the morning, for taking only three cycles of snooze rather than the usual 5 to 7.
Well done! A+!

Thank you! Yes, thank you.

A little appreciation for the everyday tasks,
is calming for the heart.

And how are you?
I mean really.
Not just as something you say to satisfy the social norms,
it has morphed from being a nicety, not a question on the side.
No, it has become hard.

What is the rhythm of your heart?
The frequency of your pain?
And is your satisfaction still louder?

Can you find a solid base of well-being?

In short: where is your happiness baseline?

Not every day of your life has to be the best ever.
(To be honest — how exhausting would life be with so much adrenalin?)
But every little pebble in your shoe would be so much easier to deal with
if your feet weren't covered in blisters from the start.
That is a fact.

Who cares how much you achieve or how much profit you can make
if you feel shit doing it and you lose all joy
and you sleep too little, and stress takes over what little you have left —

We have to stop applauding people
for being shit to themselves!

Because that is shit — and not fair towards anyone.
You have a hard time breathing, with cortisol filling your lungs.

When you are stuck in a system that forgets about people
because it tries to draw out as much wealth as it can.
And in the end, those people are told to be happy,
if it's enough for an okay life, while their fiery passion in their eyes turns cold
and it starts to burn down and it starts to burn out
all the thrills fizzle out and everywhere I look is fatigue
I see ash fill the air!

And that is why I want to wish you a very happy day of the week
from the very bottom of my heart and leave some glitter here
for being friendly to yourself
for doing the best you can on some days
for being enough as you are, without having to achieve anything
for putting your feet up and for giving no shits about capitalism
every now and again.

And for realizing that there is more than just your professional life
and that the key to equality does not lie in — and in fact can never lie in! —
trying to press everyone into the stereotypical image of the male bread winner
and people bare their teeth until permanent lockjaw.

So cheers! To a healthy dose of egoism and laughing about the doggedness,
because everyone is running too hard, I am giving a thumbs up for slowing down.

To honestly say to yourself quietly:

I am worth looking out for myself and worth liking myself.

Julien Berthier
→ S. 6 / p. 7

24/7
Arbeit zwischen Sinnstiftung und Entgrenzung

Katia Huemer

In der christlichen Tradition wurde Arbeit mit einer Strafe Gottes gleichgesetzt, die mit der Vertreibung aus dem Paradies über die Menschen gekommen war. (Wie passend, dass heute so viele unserer Arbeitsgeräte mit einem angebissenen Apfel gekennzeichnet sind.) In der antiken Gesellschaft galt körperliche Arbeit als niedere Tätigkeit, die im Widerspruch zum Ideal der philosophischen Kontemplation stand. Mit dem Aufkommen der Wissenschaften und der Überzeugung, dass Wissen er*arbeitet* werden muss, hat sich diese Sichtweise in der Neuzeit radikal gewandelt. Bis heute bestimmen das Primat der Ökonomie und das damit verbundene bürgerliche Arbeitsethos die aktuelle Gesellschaftspolitik. Die Selbst- und Fremdwahrnehmung des Individuums sowie ein gelingendes Leben hängen maßgeblich von Erwerbsbiografien ab.

Diese im Zeitraffer skizzierte historische Entwicklung stellt die heutige Arbeitswelt vor eine Vielzahl komplexer Herausforderungen und Diskussionen: Fragen der Ungleichheit, der psychischen Gesundheit am Arbeitsplatz und der Vereinbarkeit von Beruf und Privatleben bis hin zu Digitalisierung, Globalisierung und Klimawandel. Die Breite dieses

Themenspektrums widerspiegelt sich in der Ausstellung *24/7. Arbeit zwischen Sinnstiftung und Entgrenzung* in 30 künstlerischen Positionen, die sich mit unterschiedlichen Facetten dessen befassen, womit Arbeitende und Nicht-Arbeitende heute konfrontiert sind. Denn die Perspektiven auf Arbeit — für die einen bloße Notwendigkeit, für die anderen persönliche Erfüllung — haben sich in unserer Gesellschaft drastisch gewandelt. In Zeiten von permanenter Erreichbarkeit und Homeoffice sind traditionelle Arbeitszeiten längst nicht mehr die Norm. Geprägt vom Zeitalter der 24/7-Zugänge ist eine Ära scheinbarer Freiheit von starren Strukturen angebrochen. Es ist möglich, rund um die Uhr zu konsumieren — also kann (und muss) auch 24/7 gearbeitet werden. Angesichts der ständigen Begleitung durch digitale Kommunikationsmittel und des allgegenwärtigen Zwangs zur Vernetzung wird die Abgrenzung zwischen Arbeit und Nicht-Arbeit immer schwieriger. Ist das Feierabendbier schon ein privater Termin oder eher ein Meeting-Format in entspannter Atmosphäre? Bedeutet „lebenslanges Lernen“, dass wir unsere Freizeit mit Weiterbildung verbringen müssen? Und wird das Öffnen von Raum- und Zeitstrukturen vielleicht nicht nur als Befreiung, sondern auch als Überforderung empfunden?

Flexibel und entgrenzt

Flexibilität und die Fähigkeit, sich schnell und problemlos auf veränderte Anforderungen einzustellen, ist eine der vielen Kompetenzen, die in der Arbeitswelt verlangt werden. Das Verständnis des Begriffs „flexibel“, der früher durchaus positiv besetzt war, hat sich gewandelt bzw. ins Gegenteil verkehrt. Flexibilität wird vielfach nicht mehr zu beiderseitigem Vorteil ausgehandelt, sondern einseitig von den Arbeitnehmer*innen erwartet. Trotz — oder gerade wegen — der scheinbaren Freiheit von starren Strukturen und Stechkarten kämpfen viele Arbeitnehmende weiterhin um gerechte Bedingungen. Fahrradbot*innen, ehemals ein widerständiges Symbol für selbstbestimmte Arbeit und Nonkonformität, sind nun Teil der digitalen Lieferrevolution und bestimmen in knallige Farben gekleidet das Bild westlicher Städte. Die sogenannten neuen Selbstständigen (insbesondere jene, die sich in Verschuldung begeben, um Lieferdienste mit selbst erworbenen Fahrzeugen zu leisten) sehen sich existenziellen Herausforderungen gegenüber. In der Installation *Pro Line (Hi Vis — Ni Vis)*, die das Trio **Theresa**
→ S. 28 **Hattinger, Michael Hieslmair** und **Michael Zinganel** für die Ausstellung

24/7 entwickelt hat, erzählt Arbeitskleidung von jenen, die trotz ihrer auffälligen Erscheinung meist unterhalb der Wahrnehmungsschwelle städtische Infrastrukturen am Laufen halten.

Unsichtbar und ungleich

Auch andere unsichtbare Arbeit wie unbezahlte Haus- und Sorgearbeit, die immer noch überwiegend von Frauen geleistet wird, ist ein aktuelles gesellschaftliches Problem und zeigt die Notwendigkeit feministischer Perspektiven in der Arbeitsdebatte, um Ungleichheiten und Ausbeutung zu thematisieren. Der neu entstandene Essayfilm *Kein Wunder*
→ S. 62 von **Lia Sudermann** und **Simon Nagy**, der die Ausstellungen *Alles Arbeit* im Museum für Geschichte und *24/7* im Kunsthaus Graz verbindet, wirft mit Archivmaterial einen Blick aus der Vergangenheit in zeitgenössische Diskurse und schafft Momente des Vergleichs — zwischen damals und heute, aber auch zwischen den Geschlechtern. In eine
→ S. 52 ähnliche Kerbe schlug die amerikanische Künstlerin **Martha Rosler** bereits rund 30 Jahre zuvor mit *Semiotics of the Kitchen*. Die absurden bis aggressiven Gesten, mit denen sie auf Video ihr Küchen-ABC rezitiert, wirken wie eine Anleitung zum „Hausfrauenstreik". Ein feministischer Befreiungsschlag ist auch die Performance *AEG Vampyr 1400* von **Selma**
→ S. 58 **Selman**, die *dem* Symbol der Hausarbeit, dem Staubsauger, mit einer Axt zu Leibe rückt. Die symbolische Ableitung von der Axt zu Hammer und Sichel erscheint evident: Auch wenn diese beiden Werkzeuge, einst Symbole des Arbeitskampfes, zu historischen Relikten geworden sind, ist der Kampf um Gleichbehandlung, gerechte Löhne und sichere Arbeitsbedingungen heute umfassender, globaler und aktueller denn je. Mit der zunehmenden Prekarisierung der Arbeit und der Ausweitung des Niedriglohnsektors differenziert sich die Arbeitswelt immer stärker in einen Bereich von gut entlohnten und sicheren Positionen sowie einen Bereich von unterfordernden, unsicheren und existenziell bedrohlichen Beschäftigungsverhältnissen.

→ S. 12 So weist etwa **Jeremy Deller** mit dem Textilbanner *Hello, today you have day off (wording of text message sent to a worker on a zero hour contract informing him his labour would not be required that day)* auf den Zynismus hin, der unterprivilegierten Arbeitnehmer*innen leider allzu oft
→ S. 32 entgegengebracht wird. Auch **Johanna Kandl** *(Prospektor* und *Carry*
→ S. 50, 64 *On)*, **Nika Radić** *(Door — Gallery Cleaning)* und **Oliver Walker** *(One Euro)*

thematisieren in ihren jeweiligen Arbeiten soziale Ungleichheit und Einkommensunterschiede: Während Johanna Kandl in ihrer Malerei die harten Arbeitsbedingungen eines marokkanischen Minenarbeiters dem vergleichsweise angenehmen Arbeitsumfeld eines westlichen Büroangestellten gegenüberstellt, lenkt Nika Radić in ihrer Videoarbeit den Blick auf Menschen, die in unserer unmittelbaren Umgebung arbeiten und dennoch die meiste Zeit ungesehen bleiben. Oliver Walker schließlich legt den Finger in die Wunde und zeigt in seiner Videoinstallation eindrücklich die Ungleichheiten zwischen Lohn- und Zeitverhältnissen des globalen Südens und jenen des globalen Nordens
→ S. 60 auf. Auch der spanische Künstler **Santiago Sierra** thematisiert mit seiner emotional aufwühlenden Performance *Repetition of the writing of a phrase* Ungleichheiten — in seinem Fall in Bezug auf Menschen mit Migrationshintergrund und deren Chancen auf dem Arbeitsmarkt. Mit einer völlig wertfreien Haltung zeigen hingegen **Antje Ehmann**
→ S. 14 und **Harun Farocki** Videomaterial, das im Rahmen von Workshops an unterschiedlichen Orten weltweit entstanden ist. Die Präsentation des Projekts *Eine Einstellung zur Arbeit*, die die Ausstellung im Eingangsbereich einleitet, umfasst 150 Kurzfilme, die in einer einzigen Filmeinstellung verschiedenste Arbeitssituationen zeigen. Dabei spannt sich der Bogen vom Verständnis von „Arbeit" über geografische Unterschiede bis hin zu unterschiedlichen gesellschaftlichen Konzepten.

Allgemein gesprochen, werden die Veränderungen in der Arbeitswelt von mehreren Faktoren vorangetrieben. Besonders signifikant ist die Herausbildung neuer Produktionssysteme und der Aufstieg der tertiären und quartären Wirtschaftssektoren in den alten Industrieländern, die sich mehr und mehr auf Dienstleistungen und Wissenstransfer konzentrieren. Diese Entwicklung verdrängt immer mehr traditionelle landwirtschaftliche und industrielle Tätigkeiten und führt zu neuen Anforderungen und Ansprüchen an die Beschäftigten. Doch auch die Arbeitsbedingungen des primären Sektors, der auf Landwirtschaft und „Urproduktion" ausgerichtet ist, verändern sich zunehmend. **Manuel**
→ S. 10 **Correa** und **Marina Otero Verzier** richten in ihrem Video *Biological Agent* den Blick auf die sogenannte produktive Landwirtschaft, konkret auf den „Gemüsegarten Europas", der in Almería endlose Gewächshauslandschaften hervorgebracht hat. Mit routinierten Handgriffen, ausgeführt von unterbezahlten Arbeitsmigrant*innen, werden hier Obst-

und Gemüsesorten angepflanzt, gepflegt und geerntet, um schließlich überall in der EU ausgeliefert zu werden. In ihrer poetischen, tänzerischen Annäherung zollen die beiden Künstler*innen jenen Arbeiter*innen Respekt, denen Europa einen Großteil seiner Vitamine verdankt.

Selbstbestimmt und prekär

In der westlichen postfordistischen Gesellschaft, die durch Flexibilisierung, Individualisierung und Globalisierung gekennzeichnet ist, verschwimmen auch die Grenzen zwischen Selbstausbeutung und Selbstverwirklichung — insbesondere im Bereich der Kunst- und Kulturarbeit. Vor allem Künstler*innen sehen sich häufig mit Herausforderungen durch prekäre Arbeitsbedingungen konfrontiert, die von unsicheren Arbeitsverhältnissen bis unzureichender Bezahlung reichen können. Die Tatsache, dass viele Künstler*innen von Erwerbsarmut bedroht sind, wird in Werken wie *Punchcard Economy — 8 Hours*
→ S. 44 *Labour* von **Sam Meech** und *She Works Hard for the Money* von **Lisa**
→ S. 24 **Großkopf** thematisiert. Niedriglöhne (oft aufgrund von Teilzeitarbeit) und Prekarisierung sind Themen, die in den letzten Jahren verstärkt zu neuen Armutsphänomenen geführt haben; statt einer klassenlosen Gesellschaft näherzukommen, ist eine neue Klasse entstanden — die der „Working Poor". Gleichzeitig suggerieren Begriffe wie New Work, 4-Tage-Woche und Work-Life-Balance eine positive Entwicklung der Arbeitsumstände. Leere Worthülsen und Wohlfühlthemen einer digitalen Bohème oder realistische Ziele in einer Zeit höchster Belastung, die für alle Berufsgruppen gelten und die Dynamik zwischen Arbeitgeber*innen und Arbeitnehmer*innen nachhaltig verändern werden? Schließlich können wir angesichts des viel beklagten Arbeitskräftemangels frei wählen, wo, wann und unter welchen Bedingungen wir arbeiten — oder?

Autonomie ist der Leitstern unserer Kultur; wir wollen ein selbstbestimmtes Leben führen, unsere eigenen Werte und Überzeugungen und die damit verbundenen Ziele verfolgen. Selbstbestimmung statt Fremdbestimmung. *Free Agent Nation — The Future of Working for Yourself* lautet der Titel eines Bestsellers, der um die Jahrtausendwende euphorisch eine neue Ära der Innovation und Autonomie bejubelte. „It's about freedom", verkündet Amazon in der Kurzbeschreibung des Inhalts. Und weiter: „It's about time." Wahlfreiheit klingt in der Theorie

Santiago Sierra
→ S. 60 / p. 61

Sam Meech
→ S. 44 / p. 45

vielversprechend. Doch unser Anspruch, ständig nach selbstgewählten Gesetzen zu leben, macht uns nicht glücklich, sondern, wie Autonomiekritiker*innen meinen, zu selbstbezogenen Egoist*innen. Darauf verweist auch → S. 6 **Julien Berthier**s *L'horologe d'une vie de travail*, die Uhr eines Arbeitslebens, deren aufwendiges Design für genau eine Person die verbleibende Zeit bis zur Rente misst. Franco „Bifo" Berardi beschreibt in seinem Buch *Die Seele bei der Arbeit* die 1990er-Jahre als eine Zeit, in der alle Solidaritätsarchitekturen verschwunden sind, die Arbeiterklasse durch technologische Innovationen abgeschafft wurde und die Arbeit prekäre Formen angenommen hat. Tatsächlich fehlt es heute, wo jede*r für sich zu kämpfen scheint, an Massenbewegungen, die — wie die Arbeitskämpfe im Zeitalter der Industrialisierung gezeigt haben — Verbesserungen nicht nur für den Einzelnen, sondern für viele erreichen → S. 34 können. Von der Kraft der Masse erzählt auch **Peter Kogler**s Installation aus unzähligen Ameisen auf Siebdrucken, die uns auf dem Weg in die → S. 18 Ausstellung begleitet, sowie **Harun Farocki**s auf historischem Archivmaterial basierender Film *Arbeiter verlassen die Fabrik*, der die Mittel des Streiks, aber auch die „Flucht in den Feierabend" zum Inhalt hat. Auch die Bodenarbeit *The hopes and dreams of the workers as they wandered* → S. 22 *home from the bar* greift diesen Erzählstrang auf, den **Liam Gillick** in Form von Fußabdrücken in rotem Glitzersand metaphorisch andeutet.

Es sieht ganz danach aus, als seien die Hoffnungen, die die Menschen hart arbeiten ließen, um der nächsten Generation ein besseres Leben zu sichern, inzwischen verflogen. Der Mittelstand ist zum Mindeststand geworden und das neue Credo lautet: „Ich arbeite hart, damit es meinen Kindern nicht viel schlechter geht als mir." In seinem Buch *Present Shock* analysiert Douglas Rushkoff den „beständigen Strom von Dingen, Dienstleistungen und Daten", der unser Leben durchzieht. Und stellt fest: Egal wie sehr wir uns konditionieren lassen, noch mehr zu konsumieren, unseren Ressourcenverbrauch in die Höhe zu treiben und Umwelt und Menschen wirtschaftlichen Interessen zu opfern — aus ökonomischer Sicht kann es nie genug sein.

Global und innovativ

Der deutsche Soziologe Hartmut Rosa beschreibt die „Explosion der Aktivitäten", mit der die westliche Gesellschaft seit dem 18. Jahrhundert konfrontiert ist, als Modus dynamischer Stabilisierung. Er meint damit, dass das Bestehende nur erhalten werden kann, wenn wir mehr tun, wenn wir wachsen und innovativ sind – so sagt es uns zumindest unser Gefühl. Wir streben nach wirtschaftlichem Wachstum, auch wenn das, was wir bereits haben, als gut und ausreichend empfunden wird. Die Folgen dieses Denkens, das zu psychischen und physischen Krankheiten und zur Ausbeutung von Mensch und Erde führt, werden immer deutlicher. Was tun gegen den Überfluss, gegen das ständige Streben nach mehr?

→ **S. 26** Ein paar Minuten des Verweilens vor **Andreas Gursky**s großformatiger Fotografie *Amazon*, die ein menschenleeres, aber mit Waren überfülltes Lagerhaus zeigt, können in diesem Zusammenhang durchaus heilsam
→ **S. 56** sein. Auch **Christoph Schwarz** wirft in seinem Video *Supercargo – Halbautomatisches Reisen* einen humorvollen und doch ernsten Blick auf den globalen Handel, den er mit seiner eigenen Künstlerbiografie verknüpft. Um Welthandel, Finanzwesen und deren Auswirkungen auf Lebens- und Arbeitsbedingungen geht es auch in der Installation *Art,*
→ **S. 4** *Crafts and Facts* der Künstlerin **Maja Bajević**. Ihre kunstvollen Handarbeiten, die Globalisierung und menschliche Arbeit visuell eindrücklich in ein Verhältnis zueinander setzen, lassen an John Locke denken, der bereits im 17. Jahrhundert das Recht des Menschen auf Eigentum unmittelbar an „die Arbeit seines Körpers und das Werk seiner Hände" koppelte. Locke argumentierte, dass jeder Mensch einen natürlichen Besitzanspruch auf seinen eigenen Körper habe und folglich auch auf die Arbeit, die dieser Körper verrichtet. In Anlehnung an diese Lesart
→ **S. 40** kann auch **Luiza Margan**s skulpturale Installation *Cache* betrachtet werden, die auf die Wände von drei aufgeschnittenen Spinden mit Handcreme die Worte „Miete", „Lohn" und „Eigentum" geschrieben hat. Die Tatsache, dass die Creme vom Metall nach und nach absorbiert wird und die Begriffe nach einigen Wochen wieder neu angebracht werden müssen, veranschaulicht die Notwendigkeit, für existenzielle Lebensgrundlagen und – im besten Fall – Eigentum unaufhaltsam zu *arbeiten*. Denn – um erneut Locke zu strapazieren – einzig individuelle Anstrengung und Arbeit lassen privates Eigentum entstehen und bilden damit die Grundlage für eine funktionierende Gesellschaft. Oder etwa nicht?

Getaktet und kontrolliert

Die Frage nach der Zeit zieht sich als Metaebene durch die gesamte Ausstellung, worauf bereits der Titel *24/7* (die gängige Abkürzung für 24 Stunden am Tag, 7 Tage die Woche) hindeutet. Trotz des wachsenden Strebens nach Autonomie und Flexibilität fehlt vielen Menschen die Kontrolle über ihre Zeit. Denn mehr noch als Geld ist es die Verfügung über Zeit, die über Machtstrukturen entscheidet. Die Stechuhr, eine Erfindung des 19. Jahrhunderts zur Effizienzsteigerung der Arbeit, ist zu Recht ein Symbol für Kontrolle und Überwachung am Arbeitsplatz.
→ S. 30 Der taiwanesisch-US-amerikanische Künstler **Tehching Hsieh** treibt die erzwungene Taktung auf die Spitze, indem er ein Jahr lang jeweils zur vollen Stunde eine Stempeluhr betätigt hat. Ein Akt der Selbstkontrolle, aber auch eine Kritik an der auf Disziplin, Erfolg und Wachstum ausgerichteten Massengesellschaft.

Zeit ist Geld, so lautet die berühmte Maxime von Benjamin Franklin, Gründervater der USA, der sich bereits Mitte des 18. Jahrhunderts (unter anderem) Gedanken über eine bessere, d. h. effizientere Nutzung der Tageszeit gemacht hat. Aber gilt dieses Motto auch heute noch, wenn der in der Leistungsgesellschaft weit verbreitete Glaube, man lebe, um zu arbeiten, durch die veränderten Erwartungshaltungen einer neuen Generation von Arbeitnehmer*innen zunehmend ins Wanken gerät? Doch auch die auf den ersten Blick verlockende Umkehrung — man arbeite, um zu leben — bedeutet vor allem eine Reduktion der Arbeit auf eine rein ökonomische Notwendigkeit. „Work in the future is the passion that pays for itself", so lautete bereits Mitte der 1980er-Jahre der Ausgangspunkt der New-Work-Utopie von Charles Handy. „We won't be rich, but we'll be free", ergänzte der Wirtschaftsphilosoph Jahre später in einem Interview.

Arbeit ist aber nicht nur unter wirtschaftlichen Gesichtspunkten zu sehen. Sie kann und sollte auch eine Quelle von Sinn und Erfüllung im Leben eines Menschen sein, unsere Identität und unser Selbstverständnis prägen. Hannah Arendt stellte fest, dass wir der Spezies des arbeitenden Tieres *(Animal laborans)* angehören. In ihrem Werk *Vita activa oder Vom tätigen Leben* aus dem Jahr 1958 beschreibt sie die Tätigkeit des Arbeitens als zyklische und notwendige Aktivität, die durch Wiederholung und Vergänglichkeit gekennzeichnet ist.

→ S. 46 Der Künstler **Michail Michailov** macht diese Idee des Zyklischen in seiner fortlaufenden Videoserie *Just keep on going* sinnlich erfahrbar. Die kreisförmigen, wiederkehrenden Tätigkeiten, die Michailov ausführt, beziehen sich nicht zuletzt auf die grundlegende Bedingung menschlicher Existenz, die Arbeit für das Überleben des Individuums und der Gesellschaft sichert. Allerdings sieht Arendt auch die Gefahr, dass das Arbeiten den Menschen auf bloße Lebensnotwendigkeiten reduziert und ihn in einem Kreislauf der Bedürfnisbefriedigung gefangen hält. In der modernen Gesellschaft, so Arendt, könnte der arbeitende Mensch seine Fähigkeit zur Sinnstiftung und zur Teilnahme am öffentlichen und politischen Leben verlieren, wenn das Arbeiten zum dominierenden Lebensinhalt wird.

Befreit und produktiv

Der Traum von der Befreiung der Arbeit von ökonomischen Zwängen, der in solchen Überlegungen anklingt, wurde von dem deutschen Philosophen und Sozialtheoretiker Herbert Marcuse in der zweiten Hälfte des 20. Jahrhunderts geprägt. Während in einer Konsumgesellschaft, in der die Menschen auf materiellen Besitz fokussiert sind, die Arbeit zu einem bloßen Mittel zum Zweck wird, um den Konsum zu ermöglichen (was wiederum zu einem oberflächlichen und entfremdeten Lebensstil führt), könnte eine Gesellschaft, in der die Arbeit nicht mehr das zentrale Element des Lebens ist, den Menschen mehr Zeit für persönliche Entfaltung, kulturelle Aktivitäten und kreativen Ausdruck lassen. Der Ruf nach dem notwendigen Wertewandel, um soziale und ökologische Probleme in den Griff zu bekommen, wird immer drängender. Statt „Zeit ist Geld" könnte die neue Formel lauten: „Zeit ist die Voraussetzung für ein gutes Leben".

Wer reich sein will, muss Zeit verschwenden. Auf diese Neudefini-
→ S. 36 tion von Reichtum macht das Künstler*innenduo **KURS** (Mirjana Radovanović und Miloš Miletić) in seinem über die Ausstellungsdauer wachsenden Wandbild *We have always received something in exchange that we lived. On laziness* aufmerksam. Arbeit ist zweifellos eine wichtige Quelle der Resonanz, doch es kommt auf die Balance an. Auch wenn man die Vergangenheit nicht nostalgisch verklären sollte — Unterdrückung und harte Arbeit gab es seit Anbeginn der Menschheitsgeschichte —, so kam doch auch in früheren Zeiten der Moment, an

KURS (Mirjana Radovanović)
→ S. 36 / p. 37

dem „die Arbeit getan war“ und jener Zustand eintrat, in dem es keine Erwartungen mehr gab, weder von außen noch von sich selbst. Heute wird Freizeit meist für „Produktives“ genutzt. Wir „müssen“ noch Sport machen, Sozialkontakte pflegen, eine neue Sprache lernen oder wenigstens durch Social-Media-Posts scrollen, um am Laufenden zu bleiben. Ist die Zeit vergeudet, wenn sie nicht produktiv genützt wird? Führt Nichtstun automatisch zu Dekadenz oder vielleicht doch zu jenem Freiraum, der für das Entstehen kreativer Ideen so wichtig ist?

→ S. 38 *Shouldn't you be working?*, fragt **Silvio Lorusso** in seinem künstlerischen Beitrag zur Ausstellung und bezieht sich damit auf eine digitale App, die uns scheinbar vor dem Prokrastinieren schützen soll. **Elisa**
→ S. 48 **Giardina Papa** stellt mit ihrer Videoinstallation *Labour of Sleep, Have you been able to change your habits??* eine ähnliche, auf Selbstoptimierung abzielende Frage und nimmt dabei den einzigen, bis vor Kurzem noch *un*bearbeiteten Moment des Lebens in den Blick: unseren Schlaf. In seinem Buch *24/7— Schlaflos im Spätkapitalismus* argumentiert der US-amerikanische Kulturtheoretiker Jonathan Crary, dass die heutige kapitalistische Gesellschaft darauf abzielt, den gesamten Tagesablauf der Menschen zu kommerzialisieren und auszubeuten. Die ständige Verfügbarkeit von Waren und Dienstleistungen, unterstützt durch Technologie und digitale Medien, untergräbt das traditionelle Konzept von Tag und Nacht und den natürlichen Schlaf-Wach-Rhythmus des Menschen — mit schädlichen Auswirkungen auf das individuelle und kollektive Leben, da dies zu Schlafentzug, Stress und einer allgemeinen Entfremdung vom natürlichen Lebensrhythmus führt. Kurz — 24/7 macht uns kaputt.

Angesichts all dieser komplexen Veränderungen in der Arbeitswelt, die unsere Gegenwart prägen, stellt sich die Frage nach den Grenzen menschlicher Anpassungsfähigkeit. In einer nicht zu fernen Zukunft werden Technologien wie KI und eine noch weiter fortschreitende Automatisierung vieler Aufgaben die Arbeitswelt weiter verändern und Herausforderungen mit sich bringen, die erneut einen sozialen und politischen Diskurs erfordern. Doch steht die Arbeit wirklich an der Schwelle zur Immaterialität oder besteht nicht vielmehr die Gefahr, dass die körperliche, geistige und emotionale Arbeit, die mit unzähligen Bereichen unseres Lebens verbunden ist, in der KI-Euphorie vergessen wird? So ist etwa die zutiefst menschliche Fähigkeit, mit anderen in Beziehung zu treten, das Thema der Dauerperformance → S. 20 *Museum Encounters (On Work)* von **Aldo Giannotti**. Auf einer Mischung aus Podest und Bühne präsentiert der Künstler Mitarbeiter*innen des Museums, die sich bereit erklären, mit Ausstellungsbesucher*innen über ihre jeweilige berufliche Tätigkeit zu sprechen. Beziehungsarbeit inklusive.

Maschinell und effizient

Dieses kurze, performative Heraustreten aus dem Hamsterrad täglicher Routinen bringt eine unerwartete Erkenntnis mit sich: Die Welt begegnet uns als große To-do-Liste, die nie endet, sondern uns am Ende des Tages mit dem Gefühl zurücklässt, schon wieder nicht alles geschafft zu haben. Einmal mehr ist es die Zeit, von der wir immer weniger zu haben glauben. Auch wenn uns Technologien Zeit sparen (denn zweifellos lässt es sich heute schneller kommunizieren, produzieren und transportieren als noch vor einigen Jahrzehnten), übersteigt das Volumen dessen, von dem wir denken, dass wir es tun *müssen*, den Zeitgewinn. Sorgt die fortschreitende Technisierung also paradoxerweise dafür, dass wir mehr arbeiten, oder kann harte Arbeit durch Körper und Hände möglicherweise bald gänzlich von Maschinen übernommen werden? Und wenn dem so ist – sind wir an einem Punkt, an dem Marcuses Vision Wirklichkeit werden könnte und wir uns endlich mehr Zeit für Kreativität und persönliche Entfaltung nehmen dürfen? Einen Selbstversuch, der darauf abzielt, sich durch eine Maschine ersetzen zu → S. 8 lassen, stellt die Künstlerin **Louisa Clement** mit ihrer KI-gesteuerten *Repräsentantin* an. Die Puppe, die mit körperlichen und – angeblich – inneren Merkmalen der Künstlerin ausgestattet ist, steht in der Ausstellung Besucher*innen Rede und Antwort, zumindest dann, wenn ihre technischen Unzulänglichkeiten (die KI wurde 2021 programmiert und ist schlichtweg veraltet) es zulassen. Während Louisa Clement also die künstliche Intelligenz für sich arbeiten lässt, beweisen **Sebastian** → S. 54 **Schmieg** und **Silvio Lorusso**, dass die Hilfestellung zwischen Mensch und Maschine in beide Richtungen funktioniert. Mit *Five Years of Captured Captchas* feiern die beiden ein unfreiwilliges Jubiläum: fünf Jahre erfolgreiche Zusammenarbeit mit dem Konzern Google, der seine Nut- → S. 42 zer*innen für unbezahlte Mikroarbeit einsetzt. Auch **Pia Mayrwöger** präsentiert in ihrer Installation eine Maschine, genauer gesagt ein

MUSEUM ENCOUNTERS (ON WORK)
EVERY DAY BETWEEN 2PM AND 3PM A DIFFERENT WORKER OF THIS MUSEUM WILL BE PRESENT HERE AND AVAILABLE TO TALK ABOUT THEIR JOB INCLUDING HIGHLIGHTS, STRUGGLES, CONDITIONS, WISHES, FEARS, CHALLENGES BENEFITS, DOWNSIDES AND SO ON. THIS PERFORMANCE ENDS WHEN ALL THE WORKERS FROM THIS INSTITUTION HAVE PARTICIPATED.

MUSEUM ENCOUNTERS (ON WORK)
EVERY DAY BETWEEN 2PM AND 3PM A DIFFERENT WORKER OF THIS MUSEUM WILL BE PRESENT HERE AND AVAILABLE TO TALK ABOUT THEIR JOB INCLUDING HIGHLIGHTS, STRUGGLES, CONDITIONS, WISHES, FEARS, CHALLENGES BENEFITS, DOWNSIDES AND SO ON. THIS PERFORMANCE ENDS WHEN ALL THE WORKERS FROM THIS INSTITUTION HAVE PARTICIPATED.
THE MUSEUM STAFF IS TAKING A BREAK AND WILL BE THERE FOR YOU AGAIN FROM THE 10TH OF SEPTEMBER

Baugerät, als zentrales Element: Eine *Mischmaschine*, so der Titel der Arbeit, ist ihrer ursprünglichen Funktion enthoben und steckt im selbst gefertigten Baumaterial fest. Dies evoziert augenblicklich Gefühle des Mitleids mit der leblosen Arbeitshelferin. Denn die Überlegung, welchen Wert eine Maschine noch hat, wenn sie nicht mehr funktioniert, lässt sich ohne weiteres auf die menschliche Arbeitskraft übertragen.

So sehr wir uns in unserer Arbeit der Maschine annähern, so wenig sind wir intellektuell und emotional in der Lage, menschliche Handlungen so zu synchronisieren, wie es ein Computer kann. Wir brauchen uns auch nicht über Folgen wie Depressionen, Stress und ein weit verbreitetes Gefühl der Orientierungslosigkeit zu wundern, wenn wir unser soziales und berufliches Leben vollständig in die digitale Multitasking-Parallelwelt verlagern. Der Versuch, unseren Rhythmus an den der Maschinen anzupassen, hat einen hohen Preis. Wir verlieren die Fähigkeit, Dinge konzentriert zu betrachten und kreative Ideen zu kombinieren — ein Talent, das wir den Maschinen — noch — voraushaben.

Vielleicht kehren wir ja — dank oder trotz technologischer Entwicklungen — langsam wieder an den Punkt zurück, an dem die griechische Philosophie ansetzte und die Kontemplation wieder an Bedeutung gewinnen könnte. Voraussetzung dafür wäre allerdings, den nichtkommerzialisierten Bereichen der Sharing Economy eine echte Chance zu geben und sich von den Gesetzen des Geldes zu verabschieden. Oder aber wir konkurrieren weiter mit der Maschine und lassen uns — bevor das von Jeremy Rifkin und anderen Theoretiker*innen seit nunmehr 30 Jahren verkündete „Ende der Arbeit" kommt — noch einmal so richtig ausbeuten. Wir haben es in der Hand.

←← Aldo Giannotti
→ S. 20 p. 21

Literatur

Hannah Arendt, *Vita activa oder Vom täglichen Leben*, Stuttgart 1960. (Originalausgabe: *The Human Condition*, Chicago/London 1958.)

Franco „Bifo" Berardi, *Die Seele bei der Arbeit. Von der Entfremdung zur Autonomie*, Berlin 2019. (Originalausgabe: *The Soul at Work. From Alienation to Autonomy*, Los Angeles 2009.)

Jonathan Crary, *24/7. Schlaflos im Spätkapitalismus*, Berlin 2021. (Originalausgabe: 24/7. *Late Capitalism and the Ends of Sleep*, London 2014.)

Charles Handy, *The Future of Work*, New Jersey 1984.

Anja Lemke, Alexander Weinstock (Hg.), *Kunst und Arbeit. Zum Verhältnis von Ästhetik und Arbeitsanthropologie vom 18. Jahrhundert bis zur Gegenwart*, München 2014.

John Locke, *Zwei Abhandlungen über die Regierung*, hg. v. Hans Jörn Hoffmann, Frankfurt/Main 1977. (Originalausgabe: *Two Treatises of Government*, London 1689.)

Herbert Marcuse, *Das Ende der Utopie*, Berlin 1967.

Daniel H. Pink, *Free Agent Nation. The Future of Working for Yourself*, New York 2001.

Jeremy Rifking, *Das Ende der Arbeit und ihre Zukunft*, Frankfurt 1995. (Originalausgabe: *The End of Work. The Decline of the Global Labor Force and the Dawn of the Post-Market Era*, New York 1995.)

Hartmut Rosa, *Beschleunigung und Entfremdung. Entwurf einer kritischen Theorie spätmoderner Zeitlichkeit*, Berlin 2013.

Douglas Rushkoff, *Present Shock. Wenn alles jetzt passiert*, Freiburg 2014. (Originalausgabe: *Present Shock. When Everything Happens Now*, New York 2013).

Friederike Sigler, *Arbeit sichtbar machen. Strategien und Ziele in der Kunst seit 1970*, München 2021.

Oliver Walker
→ S. 64 / p. 65

Christoph Schwarz
→ S. 56 / p. 57

Louisa Clement
→ S. 8 / p. 9

the empty containers out of my head.

Sebastian Schmieg & Silvio Lorusso
→ S. 54 / p. 55

Sebastian Schmieg & Silvio Lorusso
2 Dec 2016 16:31:02 CET
To: captchas@google.com, work@google.com, collaboration@google.com, cognitive.resources@google.com, human.resources@google.com, humans@google.com, partners@google.com, info@google.com
Five Years Of Captured Captchas
Dear Google,
five years ago today, while solving yet another CAPTCHA, we wondered how much time we invest in proving that we are humans and not robots. Since then, we have archived every single CAPTCHA that we passed.
We noticed that each time we are not just proving to be human but are in fact digitising books, deciphering house numbers, and training artificial intelligence. Most of this work over the years has been done for you.
We were never asked to join your workforce. Nonetheless, we proudly accepted the task to contribute our cognitive resources to help Google achieve its goals. Today, after five years, we have collected 471 CAPTCHAs already.
Now, we are writing to present our quinquennial report which includes all the CAPTCHAs collected until now. We would be very happy to celebrate our fruitful collaboration, perhaps in your offices in Brussels or any other location you consider appropriate.
Thank you.
Best regards,
Silvio Lorusso & Sebastian Schmieg

Julien Berthier
→ S. 6 / p. 7

Johanna Kandl
→ S. 32 / p. 33

Maja Bajević
→ S. 4 / p. 5

"Unrestricted

24/7
Work between meaning and imbalance

Katia Huemer

In the Christian tradition, work was equated with God's punishment brought upon humankind with the expulsion from paradise. (How fitting that so many of our work tools today bear the logo of an apple with a bite out of it). In ancient society, physical labour was seen as a menial task that was at odds with the ideal of philosophical contemplation. With the rise of science and the conviction that knowledge must be *worked* for, this view changed radically in modern times. To this day, the primacy of the economy and the related bourgeois work ethic decide current social policy. The individual's perception of themselves and others, as well as a successful life, depend largely on employment histories.

Briefly sketched here, this historical development presents today's world of work with a host of complex challenges and discussions: the issues range from inequality, mental health in the workplace and work-life balance through to digitalisation, globalisation and climate change. The exhibition *24/7. Work between meaning and imbalance* reflects the breadth of this range of topics with 30 artistic positions exploring different facets of the conditions faced by working and non-working people today. Perspectives on work — for some a sheer necessity, for others personal fulfilment — have shifted drastically in our society. In the age of constant availability and working from home, traditional working hours are no longer the norm. Defined by the times of 24/7 access, an era of apparent freedom from rigid structures has dawned. We can consume around the clock — and so we can (and have to) work 24/7. Because of our constant connectivity to digital devices and the all-pervasive compulsion to network, the separation of work and downtime is becoming more and more difficult. Is that after-work beer actually a private drink, or is it more of a meeting format within a relaxed atmosphere? Does 'lifelong learning' mean that we have to spend our free time on further education? And is the opening up of space and time structures perhaps perceived not just as a liberation, but also as an excessive burden?

Flexible and unlimited

Flexibility and the capacity to adapt quickly and smoothly to changing requirements is one of the many skills expected in the world of work. Our understanding of the term 'flexible', which once held positive connotations, has now shifted or even reversed. In many cases, flexibility is no longer something that is negotiated to the advantage of both sides, but has become a unilateral expectation placed on employees. In spite of — or perhaps even because of — the apparent freedom from rigid structures and time punch cards, many employees are still struggling to achieve fair working conditions. Once a symbol of resistance, self-determined work and non-conformity, bicycle couriers are now part of the digital delivery revolution, shaping the image of the western city in their brightly coloured shirts. Now referred to as the 'new self-employed', these workers face existential challenges (especially those who go into debt to provide delivery services with vehicles they have bought themselves). In the installation *Pro Line (Hi Vis - Ni Vis)*, developed for the *24/7* exhibition by the trio **Theresa Hattinger, Michael Hieslmair**
→ p. 29 and **Michael Zinganel**, work clothes tell the story of those who, despite their eye-catching appearance, usually keep urban infrastructures running below the threshold of perception.

Invisible and unequal

Other invisible work — such as the unpaid housework and care still mainly done by women — is also a current social problem and shows that we need feminist perspectives in the labour debate to highlight inequalities and exploitation. *No wonder* by **Lia Sudermann** and → p. 63 **Simon Nagy** creates a link between the exhibitions *It's All Work* at the History Museum and *24/7* at the Kunsthaus Graz. The newly produced essay film explores archive material, taking a look from the past into contemporary discourse and offering moments of comparison — both between then and now and between the genders. *Semiotics of the Kitchen* by the American artist → p. 53 **Martha Rosler** struck a similar chord about 30 years previously. In her video she recites her kitchen ABCs with gestures ranging from the absurd to aggressive, like instructions for a 'housewives' strike'. → p. 59 **Selma Selman**'s performance *AEG Vampyr 1400* uses an axe to take on *the* symbol of housework, the vacuum cleaner, in a feminist act of liberation. The symbolic shift from the axe to the hammer and sickle seems evident: even if these two tools, once symbols of the labour struggle, have now become historical relics, the struggle for equal treatment, fair wages and safe working conditions is now more comprehensive, global and current than ever. With the growing precarisation of jobs and a widening low-wage sector, the world of work is increasingly splitting into two different parts: one area of well-paid, secure positions and another area of underchallenging, insecure and existentially threatening employment.

→ p. 13 **Jeremy Deller**'s textile banner *Hello, today you have day off (wording of text message sent to a worker on a zero hour contract informing him his labour would not be required that day)* highlights the cynicism that is all too often directed at underprivileged workers. → p. 33 **Johanna Kandl** (*Prospector* and *Carry On*), → p. 51 **Nika Radić** (*Door — Gallery Cleaning*) and → p. 65 **Oliver Walker** (*One Euro*) also investigate social inequality and income disparities in their respective works: while Johanna Kandl's painting contrasts the tough working conditions of a Moroccan miner with the relatively pleasant working environment of a western white-collar worker, Nika Radić's video draws attention to people who work in our immediate surroundings and yet remain unseen most of the time. Finally, Oliver Walker's video installation starkly demonstrates the painful inequalities between wage and time relations in the global south and the global north. Spanish artist **Santiago Sierra** also explores inequalities in his emotionally stirring performance *Repetition of the writing of a phrase* — in his case in relation to people with a migration background and their chances on the job market. In contrast, **Antje Ehmann** and **Harun Farocki** take an entirely neutral stance, showing video material produced during workshops at different locations around the world. The presentation of the project *Labour in a single shot,* which introduces the exhibition in the entrance area, consists of 150 short films showing all kinds of work situations in a single film shot. The themes explored range from the understanding of 'work' to geographical differences and varying social concepts.

→ p. 61

→ p. 15

Generally speaking, the changes we are currently experiencing in the world of work are driven by a number of factors. Of particularly significance is the emergence of new production systems and the rise of the tertiary and quaternary economic sectors in the old industrialised countries, where there is a growing focus on services and knowledge transfer. More and more, this development is replacing traditional agricultural and industrial activities and leading to new demands and requirements for employees. And yet there are also a number of changes to working conditions in the primary sector, which concentrates on agriculture and 'primary production'. In their video *Biological Agent,* **Manuel Correa** and **Marina Otero Verzier** investigate 'productive' agriculture and specifically the 'vegetable garden of Europe' — Almería is now home to an endless landscape of greenhouses. Varieties of fruit and vegetables are planted, tended and harvested with routine hand movements by underpaid migrant labourers, before being shipped all over the EU. Poetic and dance-like, the approach taken by the two artists pays respect to the workers who supply Europe with much of its vitamins.

→ p. 11

Self-determined and precarious

In our western post-Fordist society defined by flexibilisation, individualisation and globalisation, the boundaries between self-exploitation and self-realisation are becoming blurred — especially in the field of art and cultural work. Artists especially often face the challenges of precarious working conditions, which can range from insecure employment

Ich werde niemals einer Europäerin die Arbeit wegnehmen.
Ich werde niemals einer Europäerin die Arbeit wegnehmen.
Ich werde niemals einer Europäerin die Arbeit wegnehmen.

relations through to inadequate pay. Many artists are at risk of working poverty, a situation that is explored in works such as *Punchcard Economy — 8 Hours Labour* by **Sam Meech** and *She Works Hard for the Money*
→ p. 45
→ p. 25
by **Lisa Großkopf.** Low wages (often a consequence of part-time work) and precarisation are issues that have exacerbated new phenomena of poverty in recent years. Instead of advancing towards a classless society, a new class has emerged — the 'working poor'. At the same time, terms such as New Work, the four-day week and work-life balance point to positive developments in working conditions. Are these simply empty buzzwords and feel-good topics for a digital bohemia, or are they in fact realistic issues in an age of high stress, applicable to occupational groups across the board and set to bring about lasting changes to the dynamic between employers and employees? After all, in view of the widely lamented staff shortages, we should surely be free to choose where, when and under what conditions we work — right?

Autonomy is the guiding star of our culture. We want to lead a self-determined life, to follow our own values and convictions and the goals associated with them. Self-determination rather than heteronomy. The bestseller *Free Agent Nation — The Future of Working for Yourself* euphorically hailed a new era of innovation and autonomy at the turn of the millennium. 'It's about freedom,' proclaims the Amazon blurb. And: 'It's about time.' Freedom of choice sounds promising in theory. And yet our desire to always live according to laws of our own choosing does not make us happy but, as critics of autonomy believe, instead turns
→ p. 7
us into self-centred egoists. This is also referenced in **Julien Berthier**'s *L'horloge d'une vie de travail*, the life-work clock, whose complex design measures the time left until retirement precisely for one person. In his book *The Soul at Work*, Franco 'Bifo' Berardi describes the 1990s as a decade when the architectures of solidarity vanished, the working class was eliminated by technological innovations and work assumed precarious forms. Today, when everyone seems to be fighting purely for themselves, we do indeed see a lack of mass movements that — as shown by the labour struggles of the age of industrialisation — are capable of achieving improvements not just for the individual, but for
→ p. 35
the many. **Peter Kogler**'s installation of countless ants as screen prints, which accompanies us on our way up into the exhibition, also tells of
→ p. 19
the power of the masses, as does **Harun Farocki**'s *Workers Leaving the*

← Santiago Sierra
→ S. 60 p. 61

Factory, a film based on historical archive material that explores the methods of the strike, but also the 'escape into the free evening'. The floor work *The hopes and dreams of the workers as they wandered home*
→ p. 23
from the bar also picks up this narrative thread, with **Liam Gillick**'s metaphorical allusion taking the form of footprints in red glittering sand.

It looks very much as if the hopes that made people work hard to secure a better life for the next generation have now evaporated. The middle class has become the minimum level and the new credo is: 'I work hard so that my children are not much worse off than I am.' In his book *Present Shock*, Douglas Rushkoff analyses the 'constant flow of user rights to things, services, and data' that runs through our lives. And concludes that no matter how much we allow ourselves to be conditioned to consume even more, to accelerate our consumption of resources and sacrifice the environment and people to commercial interests — from an economic standpoint, it will never be enough.

Global and innovative

The German sociologist Hartmut Rosa describes the 'explosion of activities' that western society has experienced since the 18th century as a mode of dynamic stabilisation. What he means is that the existing situation can only be maintained if we do more, if we grow and are innovative — at least, that's what our feelings tell us. We strive for economic growth, even if what we already have is considered good and sufficient. The consequences of this way of thinking, which results in mental and physical illness and the exploitation of people and the Earth, are becoming plain to see. What can we do to counter abundance, to counter this constant striving for more?

→ p. 27
A few minutes of contemplating **Andreas Gursky**'s large-format photograph *Amazon*, which shows a deserted warehouse overflowing with
→ p. 57
goods, can be quite healing in this context. **Christoph Schwarz** also takes a humorous yet serious look at global trade in his video *Supercargo — Halbautomatisches Reisen*, which he ties in with his own artistic
→ p. 5
biography. **Maja Bajević**'s installation *Art, Crafts and Facts* is also about global trade, finance and the impact they have on living and working conditions. Her intricately crafted works create a striking visual link between globalisation and human labour, bringing to mind John

Locke, who by the 17th century had linked man's right to property directly to 'the labour of his body, and the work of his hands'. Locke argued that every human has a natural right of ownership over their own body and so also to the work that this body does.

→ p. 41 **Luiza Margan**'s sculptural installation *Cache*, with the words *Miete* (rent), *Lohn* (wages) and *Eigentum* (property) written in hand cream on the walls of three cut-open lockers, can also be seen as a reference to this reading. The cream is gradually absorbed by the metal and the words have to be reapplied every few weeks, so illustrating the need to *work* relentlessly for existential livelihoods and — at best — property. Because — in Locke's words again — only individual effort and labour can create private property and thus form the basis for a functioning society. Or does it?

Clocked and controlled

The question of time runs as a meta-level throughout the exhibition, as the title *24/7* (the common abbreviation for 24 hours a day, seven days a week) indicates. In spite of the growing aspiration to autonomy and flexibility, many people still lack control over their time. Even more than money, it is the disposal of time that determines power structures. A 19th-century invention intended to increase the efficiency of labour, it is with good reason that the time stamp clock is a symbol of control and surveillance in the workplace. The Taiwanese-American

→ p. 31 artist **Tehching Hsieh** took forced clocking in and out to the extreme by punching a time clock every hour on the hour for a year. An act of self-control, but also a critique of a mass society that is geared towards discipline, success and growth.

Time is money — the famous maxim of Benjamin Franklin, founding father of the USA, who was already thinking about a better, i.e. more efficient use of daytime by the mid-18th century (among other things). Does this motto still apply today, however, when in our achievement-oriented society the widespread belief that we live to work is increasingly being shaken by the changing expectations of a new generation of workers? But while tempting at first glance, the reverse — that we work to live — also means reducing work to a purely economic necessity. 'Work in the future is the passion that pays for itself' was the basis of Charles Handy's utopia of New Work back in the mid-1980s. '

We won't be rich, but we'll be free,' the economics philosopher added in an interview years later.

And yet work should not be seen from a purely economic perspective. It can and should also be a source of meaning and fulfilment in a person's life, shaping our identity and self-image. Hannah Arendt states that we belong to the species of working animal (animal laborans). In her 1958 work *The Human Condition,* she describes the action of work as a cyclical and necessary activity characterised by repetition and transience. The artist **Michail Michailov**'s ongoing video series *Just keep on going* makes this idea of the cyclical tangible as a sensory experience. The circular, recurring actions performed by Michailov refer not least to the fundamental condition of human existence, which ensures labour for the survival of the individual and society. However, Arendt also recognises the danger that working reduces people to mere necessities of life and traps them in a cycle of satisfying needs. According to Arendt, if work becomes the dominant purpose of life in modern society, this could lead to working people losing their ability to create meaning and be involved in public and political life.

→ p. 47

Liberated and productive

The dream of liberating work from economic constraints resonant in these reflections was shaped by the German philosopher and social theorist Herbert Marcuse in the second half of the 20th century. In a consumer society where people focus on material possessions, work simply becomes a means to an end that enables consumption (which in turn leads to a superficial, alienated lifestyle). A society where work is no longer the central element of life, however, could give people more time for personal development, cultural activities and creative expression. There is a growing urgency to the call for the shift in values that will allow us to get a grip on social and ecological problems. Instead of 'Time is money', the new formula could be: 'Time is the prerequisite for a good life'.

→ p. 37

If you want to be rich, you have to waste time. The artist duo **KURS** (Mirjana Radovanović and Miloš Miletić) highlight this redefinition of wealth in their mural *We have always received something in exchange that we lived. On laziness,* which will grow over the course of the exhibition.

KURS (Miloš Miletić)
→ S. 36 / p. 37

Work is undoubtedly an important source of resonance, and yet it depends on the balance. Although we should not nostalgically glorify the past — oppression and hard labour have existed since the beginning of human history — at least in earlier times the moment arrived when 'the work was done' and a state was achieved in which no more expectations were placed from outside or from oneself. Today, our free time is mostly used for doing something 'productive'. We still 'have to' work out, socialise, learn a new language or at least scroll through social media posts to keep up to date. Is time wasted if it is not used productively? Does doing nothing automatically lead to decadence — or perhaps instead to the freedom that is so important for coming up with creative ideas?

→ p. 39 *Shouldn't you be working?* asks **Silvio Lorusso** in his contribution to the exhibition, referring to a digital app that is supposed to protect us from procrastination. In her video installation *Labour of Sleep, Have you been* → p. 49 *able to change your habits??*, **Elisa Giardina Papa** poses a similar question aimed at self-optimisation, investigating the only moment in life that had until recently remained unprocessed: our sleep. In his book *24/7. Late Capitalism and the Ends of Sleep*, US cultural theorist Jonathan Crary argues that today's capitalist society aims to commercialise and exploit the whole of our daily routine. The constant availability of goods and services supported by technology and digital media erodes the traditional concept of day and night and the natural sleep-wake rhythm of humans — with a damaging effect on individual and collective life, since it causes sleep deprivation, stress and a general alienation from the natural rhythm of life. In short — 24/7 is destroying us.

In view of the complex changes in the world of work that shape our present-day reality, we must ask ourselves where the limits of human adaptability lie. In the not-too-distant future, technologies such as AI and the further automation of many tasks will keep on changing the world of work and bring with them challenges that again call for social and political discourse. But is work really on the brink of immateriality, or is it not more the case that there is a danger that the physical, mental and emotional labour associated with countless areas of our lives will get forgotten in the AI euphoria? The profoundly human ability to → p. 21 enter into relationships with others is the subject of **Aldo Giannotti**'s durational performance *Museum Encounters (On Work)*. On a mix of podium and stage, the artist presents museum employees who have agreed to talk to exhibition visitors about their various professional activities. Including relationship work.

Automated and efficient

The brief, performative escape from the hamster wheel of the daily routine brings with it an unexpected realisation: we encounter the world as one long to-do list that never ends and at the end of the day simply leaves us with the feeling that we haven't managed to get everything done. Once again, it is time that we find we have less and less of. Even if technology saves us time (because we can undeniably communicate, produce and transport things faster today than we did a few decades ago), the volume of what we think we *need to* do exceeds the time we save. And so, paradoxically, could it be that advancing mechanisation is making us work more, or could the hard labour done by our bodies and hands soon be taken over entirely by machines? If this is indeed the case, are we at a point where Marcuse's vision could become reality and we can finally take more time for creativity and personal fulfilment? → p. 9 The artist **Louisa Clement** is conducting a self-experiment aimed at having herself substituted by a machine in her AI-controlled *Representative*. The doll, which features the artist's physical and — allegedly — inner characteristics, will answer questions from visitors at the exhibition, at least if its technical shortcomings (the AI was programmed in 2021 and is simply outdated) allow. While Louisa Clement gets artificial intelligence to work for her, **Sebastian Schmieg** and **Silvio** → p. 55 **Lorusso** demonstrate that assistance between humans and machines works both ways. In *Five Years of Captured Captchas*, the duo celebrates an involuntary anniversary: five years of successful collaboration with → p. 43 Google, which utilises its users for unpaid micro-labour. **Pia Mayrwöger**'s installation has a machine, or more precisely construction equipment as its central element. A *Mischmaschine* (concrete mixer), as the work is called, has been stripped of its original function and is stuck in the self-made building material. This immediately evokes feelings of pity for the lifeless worker. The question as to what value a machine has if it no longer functions can, after all, easily be transferred to human labour.

While we may be moving closer to the machine in our work, we are not intellectually and emotionally capable of synchronising human actions in the same way that a computer can. Nor should consequences like depression, stress and a widespread feeling of disorientation surprise us if we choose to shift our social and professional lives entirely into the digital multitasking parallel world. The attempt to adapt our own rhythms to that of machines comes at a high price. We lose the ability to observe things closely and to combine creative ideas — a talent that we — for now — still have over machines.

Perhaps — thanks to, or in spite of, technological developments — we are slowly returning to the point where Greek philosophy began and contemplation could regain its importance. However, the prerequisite for this would be to give the non-commercialised areas of the sharing economy a real chance and to say goodbye to the laws of money. Or we can continue to compete with the machine and — before the 'end of work' declared by Jeremy Rifkin and other theorists over the last 30 years — allow ourselves to be exploited once again. It is in our hands.

Literature

Hannah Arendt, *The Human Condition*, Chicago/London 1958.

Franco 'Bifo' Berardi, *The Soul at Work. From Alienation to Autonomy*, Los Angeles 2009.

Jonathan Crary, 24/7: *Late Capitalism and the Ends of Sleep*, London 2014.

Charles Handy, *The Future of Work*, New Jersey 1984.

Anja Lemke, Alexander Weinstock (eds.), *Kunst und Arbeit. Zum Verhältnis von Ästhetik und Arbeitsanthropologie vom 18. Jahrhundert bis zur Gegenwart*, Munich 2014.

John Locke, *Two Treatises of Government*, London 1689.

Herbert Marcuse, 'The End of Utopia,' in: *Five Lectures: Psychoanalysis, Politics, and Utopia*, Boston 1970 (Original: *Das Ende der Utopie*, Berlin 1967.)

Daniel H. Pink, *Free Agent Nation. The Future of Working for Yourself*, New York 2001.

Jeremy Rifkin, *The End of Work. The Decline of the Global Labour Force and the Dawn of the Post-Market Era*, New York 1995.

Hartmut Rosa, *Beschleunigung und Entfremdung. Entwurf einer kritischen Theorie spätmoderner Zeitlichkeit*, Berlin 2013.

Douglas Rushkoff, *Present Shock. When Everything Happens Now*, New York 2013.

Friederike Sigler, *Arbeit sichtbar machen. Strategien und Ziele in der Kunst seit 1970*, Munich 2021.

Louisa Clement
→ S. 8 / p. 9

Sam Meech
→ S. 44 / p. 45

Michail Michailov
→ S. 46 / p. 47

Michail Michailov
→ S. 46 / p. 47

From beginning to the end.

Pia Mayrwöger
→ S. 42 / p. 43

Lia Sudermann & Simon Nagy
→ S. 62 / p. 63

Pia Mayrwöger
→ S. 42 / p. 43

Johanna Kandl
→ S. 32 / p. 33

Carry On!
Johanna Kandl

Unser gemeinsames Problem — oder ist Kunst Arbeit?

Katja Praznik

Gerne thematisieren Künstler*innen die kapitalistische Ausbeutung von Arbeiter*innen, wie es viele auch in dieser Ausstellung tun, aber zumeist identifizieren sie sich selbst nicht als solche, noch sehen sie ihre Situation als Ausbeutung durch das System der westlichen Kunstinstitution. Zum Großteil tragen Künstler*innen freiwillig zur Mystifikation von Arbeit und Arbeitsbedingungen in der Kunstwelt bei und reproduzieren diese oft zu ihrem eigenen wirtschaftlichen und sozialen Nachteil. Wenn der Fokus dieser Ausstellung auf den Widersprüchen von Arbeit zwischen Sinnstiftung und Entgrenzung liegt, so sind Kunstarbeiter*innen eines der besten Beispiele für diese Antinomie. Vor allem anerkannte Künstler*innen, deren Arbeiten wir in Ausstellungen und Biennalen betrachten, sind generell nicht bereit, sich öffentlich über die dunkle Seite eines Prozesses zu äußern, welcher den ästhetischen/künstlerischen Wert ihrer Arbeiten etabliert hat und den Pierre Bourdieu treffend als „Weihe" beschreibt.[1] Diese Künstler*innen machen die helle, sichtbare Seite der „kreativen, dunklen Masse" aus, ein Begriff von Gregory Sholette, welcher all jene Künstler*innen umfasst, deren Arbeiten nicht anerkannt, nicht entgolten oder anderweitig marginalisiert werden. Kurz gesagt also all jene Künstler*innen, die es nie „schaffen" werden, weil sie entweder die westliche Kulturinstitution ablehnen und beschlossen haben, in deren Schatten zu arbeiten – oder weil sie glauben, keine andere Wahl zu haben als unsichtbar zu bleiben.[2] Die strahlenden Ausnahmefälle existieren nicht trotz, sondern gerade aufgrund der Existenz dieser kreativen, dunklen Masse. Sie sind diejenigen, die in Kunstmuseen und Galerien ausgestellt werden.

Künstler*innen sind keine Opfer in diesem Chiaroscuro. Im Gegensatz zu Physiker*innen, die der Frage nachgehen, weshalb „die Realität nicht so ist, wie es scheint"[3], „ignorieren die Einwohner*innen der Welt der institutionellen Kunst [...] großteils den ungesehenen Zuwachs von Kreativität, von dem sie trotz allem abhängig bleiben"[4]. Viele von ihnen nehmen zahm an einem Spiel teil, in dem es nur wenige herausstechende, furchtlose Akteur*innen gibt, die sich öffentlich gegen Ungleichheit und Ausbeutung äußern, die in der westlichen Kunstinstitution so verankert sind. Die Freigeister, die aufbegehren, tun dies jedoch mit dem Risiko, ausgeschlossen, wie die Pest vermieden, als schwierig abgestempelt oder mit anderen abwertenden Beinamen bedacht zu werden. Darüber hinaus werden solche Aktionen oft als Teil des Nach-Aufmerksamkeit-Heischens unserer spektakelsüchtigen Gesellschaft interpretiert, obwohl die kreative, dunkle Masse leise und unsichtbar immer auf neue und wiederkehrende Mitglieder wartet. Die Macht der Kunstinstitution und deren Umfeld ist dermaßen groß, dass solche Provokationen auch oft als politisch neutralisierte Kunstwerke und harmlose Provokationen für den Kunstgenuss bereitgestellt werden.

Die Antwort auf die Frage, weshalb Künstler*innen nicht bezahlt werden wollen, könnte genauso gut sein, dass sie es sich einfach *leisten* können, gratis zu arbeiten. Die meisten Künstler*innen, die im Bereich Kunstproduktion arbeiten, tolerieren diese Arbeitsbedingungen und Spielregeln. Sie lieben, was sie tun, und tun, was sie lieben. Die Arbeit gibt diese Liebe nicht zurück, die Künstler*innen sind vielmehr in einer Position, die es ihnen erlaubt, freie Arbeit anzubieten im Austausch für Anerkennung, Sichtbarkeit, Freude und anderweitigen immateriellen und kurzweiligen Status-bringenden Ruhm, der – natürlich – in keiner Weise dazu beiträgt, die Rechnungen zu zahlen. Tatsächlich „arbeiten Künstler*innen nicht für Geld. Zumindest denken das viele Leute. Es ist eine wunderbare Ausrede, um Künstler*innen nicht für ihre Arbeit zu bezahlen“[5]. Immer wieder belegen Studien zu den Arbeitsbedingungen in der Kunstwelt, in welchem Ausmaß schlechte oder nicht existente Bezahlung des künstlerischen Arbeitsprozesses normalisiert wird und wie unsicher die wirtschaftliche Situation und Arbeitsbedingungen von Künstler*innen weltweit sind. Soziolog*innen und Künstler*innen selbst liefern solide empirische Daten[6] für die Hypothese, dass das entgeltlose Arbeiten für Künstler*innen kein Thema ist, sondern eine anerkannte Norm. Das erklärt, weshalb sich Künstler*innen selbst eher als Unternehmer*innen denn als Arbeiter*innen sehen. Wie viele Gelehrte anmerken, sind Künstler*innen die idealen Arbeiter*innen für das neoliberale, unternehmerische Modell. Die große Ideologie hinter der westlichen Kunst ist es, das Werken einer Künstlerin*eines Künstlers nicht als Form von Arbeit zu sehen, sondern als Schaffensakt, womit Künstler*innen nicht Handwerker*innen sind, sondern eine Art Kolonie von Menschen, die einer heiligen Berufung nachgehen. Das ist zweckmäßig für die neoliberale Ideologie, die Autonomie, Freiheit, Selbstständigkeit und Kreativität zelebriert. In dieser verkehrten Welt werden Sinnhaftigkeit, Freude und Liebe zur Arbeit zur Rechtfertigung für niedrige oder nicht existente Bezahlung – die Tatsache, dass wir unseren Lebensunterhalt verdienen müssen, wird dabei komplett außer Acht gelassen (zu diesem Punkt kehre ich später noch zurück). Einige Soziolog*innen bezeichnen diese Praxis als „Opferarbeit“, eine Art freiwilligen ökonomischen Rabatt für die Freude an der eigenen Arbeit.[7] Diese Perspektive vernachlässigt die Tatsache, dass die Ausführung von Opferarbeit trotzdem nur jenen möglich ist, die es sich auch leisten können.

Ich bevorzuge die Ansichtsweise der marxistischen Feminist*innen und deren bahnbrechender Kritik von unbezahlter Haus- und Fürsorgearbeit, die ich über die Idee von „unsichtbar gemachter Arbeit“[8] auf die Kunstwelt umlege. Wenn man Kunstarbeit (die Aktivität, nicht das Objekt, das Kunstwerk) als Form von unsichtbarer Arbeit betrachtet, kann man die Mechanismen erkennen, die in wirtschaftlich rabattierter und ausgebeuteter Arbeit resultieren.[9] Dadurch spricht man auch den Elefanten in Raum an, nämlich die Tatsache, dass „in einer Gesellschaft, in der monetäres Einkommen eine Voraussetzung für das Überleben ist, das Liefern von Kapital ohne jegliches Entgelt ein Rezept für garantiertes Scheitern ist“[10]. Die historische Analyse von

Apparatus 22, *ART IS WORK*, Barcelona, 2021

Kunstarbeiter*innen[11] und Hausarbeiter*innen bestätigt in beachtlichem Umfang, dass unbezahlte Arbeit Künstler*innen bislang nicht befreit hat, und zeigt ganz im Gegenteil, dass unbezahlte Arbeit Kunstarbeiter*innen in eine Position bringt, in der sie jeden Auftrag und jede Bezahlung annehmen müssen.[12] Somit liegt alle Macht im Bereich der Kunstproduktion nicht bei den Künstler*innen selbst, sondern bei jenen Akteur*innen, die institutionelle Macht in den Händen halten, die finanziellen Ressourcen in bildenden Institutionen kontrollieren, wo sie die Ideologie von Kreativität und künstlerischem Genie nähren, und bei Kunst- und Kulturorganisationen, welche diese Ideologie in ein auf unbezahlter Arbeit beruhendes Geschäftsmodell umwandeln.

Trotzdem sind nicht alle Kunstinstitutionen Teil der dunklen Seite. In diesem global normalisierten Standard des „Geschäftsmodells unbezahlter Arbeit", auf dem Museen und Galerien normalerweise beruhen, können dieses Kunsthaus und die Ausstellung *24/7* als positives Beispiel gelten. Die Künstler*innen hinter dieser Ausstellung konnten den von dem multidisziplinären Kunstkollektiv Apparatus 22 produzierten

Overall mit folgendem Aufdruck tragen: „Ich habe ein Künstler*innenhonorar in dieser Ausstellung erhalten. Ihr habt verstanden, dass Kunst Arbeit ist."[13] Jede*r Künstler*in in dieser Ausstellung erhielt ein Ausstellungshonorar, je nachdem, ob sie ein bereits existierendes Werk verliehen bzw. adaptiert oder ein neues produziert hatten. Diejenigen, die für die Ausstellung neue Werke produzierten, bekamen die Produktionskosten rückerstattet; weiters wurden Reisekosten nicht mit Honoraren verwechselt. Das Kunsthaus Graz folgt den Empfehlungen und Richtlinien der IG Bildende Kunst betreffend ein Mindestentgelt für Kunstarbeit, welche nach jahrzehntelangem Kampf von Kunstarbeiter*innen um „faire Bezahlung" in den Künsten etabliert wurden.[14] Ein Drittel des Budgets für diese Ausstellung wurde für Künstler*innenentgelte und Kunstproduktion ausgegeben. Zweifelsohne ist das ein Schritt zur Anerkennung von Kunst als Arbeit und der Tatsache, dass Künstler*innen für die Arbeit, die sie für Museen und Galerien leisten, entlohnt werden müssen. Dennoch hält sich der Großteil der Kunstinstitutionen nicht an solche Standards. Diese stützen sich auf — und propagieren — ein Geschäftsmodell, das nach wie vor auf unbezahlter Arbeit beruht, während wir, das entzückte Publikum, den Nervenkitzel herausragender Kunst genießen.

Warum ist es so schwer, Kunst als Arbeit anzuerkennen? Warum verschleiern wir den Arbeitsprozess, anstatt ihn ordentlich zu evaluieren? Jede*r Künstler*in kann bestätigen, dass ein Ausstellungsentgelt nicht den Arbeitsprozess deckt, der hinter der Produktion eines Kunstwerks steht. Auch wenn Kurator*innen und Museumsdirektor*innen, so wie in diesem Museum hier, sich der Machtdynamik bewusst sind, innerhalb derer sie agieren, sind sie doch eine Minderheit. Warum tolerieren wir diese Situation? Mindert es den Kunstgenuss, wenn Kunstarbeiter*innen für ihre Arbeit bezahlt werden? Muss der Sinn der Arbeit einer Künstlerin*eines Künstlers auf solch ungleicher wirtschaftlicher Basis beruhen? Warum reproduzieren wir diese Konditionen und machen es weniger privilegierten und wirtschaftlich nicht abgesicherten Individuen unmöglich, jemals ihren Lebensunterhalt mit Kunst zu verdienen?

Seinen Lebensunterhalt zu verdienen ist immerhin ein Problem, das wir alle gemeinsam haben. „Denn Kapitalismus beginnt nicht mit einem Arbeitsangebot, sondern mit der Notwendigkeit, seinen Lebensunterhalt zu verdienen."[15] Die Sinnstiftung oder die Entgrenzung von Arbeit und den vielfältigen Arbeitswelten von heute geben Anlass dazu, die Ungerechtigkeiten, die im vorherrschenden kapitalistischen Produktionsmodus verankert sind, zu überdenken. Karl Marx, einer der wichtigsten Denker zum Thema Anforderungen der Arbeitswelt und der Emanzipation von Arbeiter*innen unter dem Kapitalismus, beschreibt in *Das Kapital* Arbeit als „eine von allen Gesellschaftsformen unabhängige Existenzbedingung des Menschen, ewige Naturnotwendigkeit, um den Stoffwechsel zwischen Mensch und Natur, also das menschliche Leben zu vermitteln".[16] So wie Spinnen Netze weben und Bienen Honigwaben bauen, so arbeiten auch Menschen und erfüllen dadurch ihren Lebenssinn. Marx zögert jedoch nicht lange, uns auch daran zu erinnern, dass unsere Arbeit unter dem Kapitalismus die Form eines Guts, von Arbeitskraft oder Arbeitskapazität annimmt, welche nicht nur eine beliebige Anzahl an nützlichen Dingen produziert (use-values), sondern auch Wert (value) generiert.[17]

Während Marx' Theorie von Arbeit als wichtigste wertschöpfende Substanz im Kapitalismus für unser Verständnis von Arbeit von zentraler Bedeutung ist, beschränkte er sie auf den Bereich der Produktion. Die feministische Kritik an unsichtbarer reproduktiver Arbeit brachte jedoch einige wichtige und fehlende Tatsachen ans Licht. Patriarchalische, imperiale Scheuklappen hindern uns oft daran, die entgeltlose Welt zu sehen, auf der bezahlte oder entlohnte Arbeit beruht. Anders ausgedrückt, gibt es mehrere Formen der Arbeit — die wichtigste davon ist die Arbeit der gesellschaftlichen Reproduktion, welche sich auf alle Aktivitäten bezieht, die das Leben, den Lebensunterhalt und die kapitalistischen Beziehungen regenerieren — die nicht als produktiv gelten, sondern wirtschaftlich unsichtbar sind,[18] obwohl sie „die wertschöpfende Substanz des Kapitals" sind.[19] Ebenso bedeutsam ist, dass der Prozess der Wertschöpfung, für den unsere Arbeit essenziell ist, unter verschiedenen Formen der Ausbeutung stattfindet, und zwar unabhängig davon, ob man einen Lohn erhält oder nicht.[20] Vereinfacht ausgedrückt, ist im Kapitalismus keine Form der Arbeit unabhängig und davor gefeit, zu einer wertschöpfenden Substanz und folglich auch der Ausbeutung unterworfen zu werden, selbst wenn sie als Nichtarbeit erscheint.

Man könnte sagen, dass dies die Produktion von Kunst sowie öffentliche Kunst-Organisationen nicht betrifft, da diese gemeinnützig und keine kapitalistischen Konzerne sind. Die Menschen, die in und für solche Organisationen arbeiten, finden ihre Arbeit möglicherweise auch sinnstiftend und vielleicht sogar erfüllend, aber sie existieren in Strukturen, die von dem Zwang, einen Lebensunterhalt zu verdienen, dominiert und definiert sind, der uns wiederum von kapitalistischen Vernetzungen und dem kapitalistischen Durst für Anhäufung auferlegt wurde. Der Zweck von Kunst mag nicht im Profit liegen, aber heißt das gleichzeitig, dass Künstler*innen und alle anderen in der Kunstproduktion involvierten Personen von den Anforderungen dieser kapitalistischen Vernetzungen befreit sind? Wie nun agiert dieses spezielle Gut, welches „die besondere Eigenschaft besitzt, Wert zu generieren"[21], im Zusammenhang von Kunst in der kapitalistischen Gesellschaft? Wie ich bereits angemerkt habe, fasst dies der Untertitel dieser Ausstellung sehr gut zusammen. Ich darf ein weiteres Beispiel anführen: Ob ein Urinal ein wertvolles Kunstwerk ist oder nicht, hängt davon ab, ob eine Gruppe von Künstler*innen es als solches erdacht und präsentiert hat, ein Museum es ausgestellt hat, Kritiker*innen darüber geschrieben und Besucher*innen es bewundert haben. Kunst ist in der Tat eine kollektive Übung, genauso wie das Schaffen von ästhetischem Wert. Viele Personen sind involviert, aber sobald das symbolische Kapital in wirtschaftliche Parameter umgewandelt wird, sind die Künstler*innen benachteiligt. Wer profitiert (heute noch) von dem Wert des berühmtesten Urinals der Welt? Die Kunstarbeiter*innen oder die außerordentlich gut bezahlten hochrangigen Angestellten des Museums? Basierend auf recherchierten Fakten[22] ist diese Angelegenheit ein Beispiel für ein eindeutiges, schweres Ungleichgewicht. Verstehen wir Kunstschaffenden so das Ungleichgewicht, welches der Sinnhaftigkeit unserer Arbeit entspringt?

So sehr manche sehnsüchtig das Ende von Arbeit anhand von technologisch-utopischem Wandel wie Automatisierung, Robotisierung, Genmanipulation und Ähnlichem prophezeien, so hat uns die globale Covid-19-Pandemie deutlich gemacht, dass die Arbeit im Kapitalismus kein Ende hat. Es gibt seither sogar eine neue Unterscheidung zwischen „essenziellen" und „nicht-essenziellen" Arbeiter*innen und Arbeit. Es sollte nicht an uns vorübergehen, dass es eben jene als informell abgestempelten Haus- und Sorgearbeiter*innen waren, die während der Pandemie als „globale Hausarbeit" sichtbar wurden.[23] Diese Arbeiter*innen ermöglichen den Kapitalismus, nicht trotz ihrer entgeltlosen Existenz, sondern gerade deswegen. Während das Ungleichgewicht in Sachen Arbeit im Großen und Ganzen auf der entgeltlosen, informellen Arbeit eines Großteils der Weltbevölkerung beruht, wollen wir trotzdem zu unserer kleinen, wenn auch globalen Kunstwelt zurückkehren und uns mit dem Geschäftsmodell von unentgoltener Arbeit und den unzähligen unbezahlten Stunden von Kunstarbeiter*innen befassen, welche in der Tat die größte Subvention darstellen, die Kunstinstitutionen immer noch erhalten.

Trotz des Mächte-Konsortiums der globalen Welt der Kunstarbeit (oder besser, des globalen Kunst-Arbeitsmarktes) sollte man die Macht, die wir als Kollektiv — zu kritischem Denken und Handeln fähig — in den Händen halten, nicht unterschätzen. Es ist richtig, dass wir individuell als Wettstreiter*innen im Arbeitsmarkt wenig Einfluss haben, aber wenn wir beginnen, in Solidarität die Ausbeutung unbezahlter Arbeit von unserer Machtposition aus — wenn sie auch noch so klein ist — zu entnormalisieren, dann können wir die Machtkonstellation kollektiv verändern. Ich bin nach wie vor überzeugt davon, dass die Anerkennung von Künstler*innen als Arbeiter*innen und von Kunst als einer Form von Arbeit, wie Hausarbeit oder jede andere Form von Arbeit, die Wert generiert und zur kapitalistischen Akkumulation beiträgt, ein notwendiger Schritt ist, wenn wir damit aufhören möchten, dem Kapital kostenlose Geschenke zu bereiten und stattdessen die Kunst als Schauplatz des Kampfes für eine Lebensweise und Wirtschaft sehen, die auf dem Wert von menschlichen Bedürfnissen und Arbeit basieren, anstatt auf der Wertschöpfung im Namen des Profits.

Fußnoten

1 Vgl. Pierre Bourdieu, *Rules of Art: Genesis and Structure of the Literary Field*, Stanford 1992.
2 Vgl. Gregory Sholette, *Dark Matter: Art and Politics in the Age of Enterprise Culture*, New York 2011.
3 Carlo Rovelli, *Reality Is Not What It Seems: The Journey to Quantum Gravity*, New York 2017.
4 Sholette 2011.
5 Dieter Lesage, „A Portrait of the Artist as a Worker", in: Maska 20, Nr. 94—95, 2005, S. 93 f., hier S. 93.
6 Vgl. Jaka Primorac, Valerija Barada, Edgar Buršić, „Creative Workers in Permanent Crisis: Labor in Croatia's Contemporary Arts and Culture", in: Ilya Kiriya, Panos Kompatsiaris, Yiannis Mylonas (Hg.), *Industrialization of Creativity and Its Limits*, Cham 2020; Vahida Ramujkić, Milan Đorđević (Hg.), *Rad u umetnosti. Zbornik udruženja likovnih umetnika Srbije. Belgrad: Udruženje likovnih umetnika Srbije.* [Work in Art: An Edited Volume of the Association of Fine Artists of Serbia], Belgrad 2021.
7 Vgl. Andrew Ross, „The Mental Labor Problem", in: Social Text 18, Nr. 2, 2000, S. 1—31.
8 Katja Praznik, „Lexicon: Invisiblized Labor", in: Noah Brehmer, Vaida Stepanovaite, *Paths to Autonomy*, Brooklyn 2022.
9 Vgl. Katja Praznik, *Art Work: Invisible Labour and the Legacy of Yugoslav Socialism*, Toronto 2021.
10 Silvia Federici, *Patriarchy of the Wage: Notes on Marx, Gender and Feminism*, Oakland 2021.
11 Vgl. Praznik 2021.
12 Federici 2021, S. 10.
13 NICC Vitrine, *Apparatus 22 — „Art is Work"*, 25. Februar 2022, in: NICC, https://nicc.be/project/apparatus-22-art-is-work/ (14.06.2024).
14 IG Bildende Kunst, *Leitfaden für faire Bezahlung in der bildenden Kunst*, Wien 2024, https://igbildendekunst.at/infomaterial/honorare/ (14.06.2024).
15 Michael Denning, „Wageless Life", in: New Left Review 66 (Nov—Dez) 2010, S. 79—97.
16 Karl Marx, *Das Kapital*, Bd. 1: *Kritik der politischen Ökonomie. Der Produktionsprozeß des Kapitals*, Hamburg 1850, S. 283.
17 Vgl. ebda.
18 Vgl. Mariarosa Dalla Costa, Selma James, *The Power of Women and the Subversion of the Community*, Bristol 1973; Silvia Federici, *Wages against Housework*, Bristol 1975; Maria Mies, *Patriarchy and Accumulation on a World Scale*, London 1986.
19 Alessandra Mezzadri, „Value theories in motion: Circular labour migration, unfinished land dispossession and reproductive struggles across the urban—rural divide", in: Environment and Planning F 0, Nr. 0 (26. Jänner 2024), https://doi.org/10.1177/26349825231224027 (14.06.2024).
20 Vgl. Jairus Banaji, *Theory as History: Essays on Modes of Production and Exploitation*, Leiden 2010; Mezzadri, 2024.
21 Marx 1850, S. 270.
22 Philip Boroff, „Museum of Modern Art Reveals Glenn Lowry's Whopping $2.1 Million Pay Amid Staff Protest Over Benefit Cuts", in: Artnet, 4. Juni 2015, https://news.artnet.com/art-world/glenn-lowrys-2-1-million-package-revealed-304820 (14.06.2024).
23 Alessandra Mezzadri, „The Social Reproduction of Pandemic Surplus Populations and Global Development Narratives on Inequality and Informal Labour", in: Development and Change 53, Nr. 6, 2022, S. 1230—1253.

Liam Gillick →
→ S. 22 p. 23

Selma Selman
AEG Vampyr 1400, 2017
Dokumentation der Performance im acb Attachment, Budapest, Video, 9:46 min, Staubsauger Documentation of the performance at acb Attachment, Budapest, video, 9:46 min, vacuum cleaners

Selma Selman
→ S. 58 / p. 59

Santiago Sierra
→ S. 60 / p. 61

Nika Radić ↗
→ S. 50 / p. 51

Pia Mayrwöger →
→ S. 42 / p. 43

Nika Radić
→ S. 50 / p. 51

Pia Mayrwöger
→ S. 42 / p. 43

Aldo Giannotti
→ S. 20 / p. 21

Aldo Giannotti
→ S. 20 / p. 21

ERS (ON WORK)
PM AND 3PM A
THIS MUSEUM WILL
VAILABLE TO TALK
NG HIGHLIGHTS,
SHES, FEARS, CHALLENGES
D SO ON. THIS
N ALL THE WORKERS
HAVE PART TED.

Our mutual predicament, or is art work?

Katja Praznik

Artists like to talk critically about the capitalist exploitation of other workers, as many do in this exhibition, but they mostly tend not to identify as workers themselves, nor perceive their situation as exploited by the system of the Western institution of art. For the most part, artists freely participate and reproduce the mystifications about labor and working conditions in the arts — often to their own economic and social detriment. If this exhibition is centered on the contradictions of work between meaning and imbalance, then art workers are one of the best embodiments of this antinomy. Especially recognized artists whose works we see in exhibitions and biennales are not prone to talk publicly about the dark side of the process through which the aesthetic/artistic value of their work has been established, and which Pierre Bourdieu astutely describes as 'consecration'.[1] These artists are the bright visible side of the 'creative dark matter', a term coined by Gregory Sholette to denote all the artists whose work goes unrecognized, uncompensated, or marginalized. In short, all those artists who will never 'make it' either because they reject the Western institution of art and have decided to work in its shadows, or, mostly, because they think they don't have any other choice but to remain invisible.[2] Those exceptions who shine do so not in spite of, but *because of* the creative dark matter. They are the ones who are put on the walls and in the halls of art galleries and museums.

Artists are not victims in this chiaroscuro. Contrary to physicists who explore why 'reality is not what it seems',[3] in the realm of institutionalized art its 'denizens... largely ignore the unseen accretion of creativity they nevertheless remain dependent on.'[4] Most of them docilely participate in the game with a few dazzling courageous offenders who decide to speak up publicly against the inequality and exploitation that is inscribed in the Western institution of art. However, the free spirits who speak up do so at the risk of being ostracized, avoided like the plague, labelled as difficult, and showered with other deprecatory epithets. Even more, their actions are often taken as part of the recognition game in our society of spectacle, although the creative dark matter is there silently and invisibly welcoming new or returning members. The power of the institution of art and its coterie is such that provocations might also end up becoming artworks neutralized politically and ready for art consumption as benign provocations.

The answer to the question of why artists don't want to get paid might as well be because they simply can *afford* to work for free. Most artists who work in the field of art production tolerate these working conditions and rules of the game. They love what they do and do what they love. The work doesn't love them back; rather, they simply are in a position that allows them to offer free labor on account of recognition, exposure, enjoyment, and other immaterial and ephemeral status-gaining glory, which, needless to say, doesn't pay the bills. Indeed, artists are 'not doing it for the money. That is what some people think. It is a great excuse not to pay the artist for the work they do'.[5] Time and again, sociological research about working conditions in the arts demonstrates how normalized nonpayment or poor payment for artists' actual labor process is, and how insecure artists' economic and labor conditions are on a global scale. Sociologists and artists themselves provide solid empirical data[6] for a hypothesis that working for free is not an issue for artists, but rather an accepted norm. This explains why artists prefer to consider themselves as entrepreneurs rather than workers. As many scholars note, artists are the model workers for neoliberal entrepreneurial rationality. The grand ideology of Western art is that what artists do is not a form of work, but rather an act of creation — artists are not artisans, but a colony of those with a sacred calling. This is expedient for the neoliberal ideology that celebrates autonomy, freedom, self-reliance, and creativity. In this upended world, the joy, love, or meaningfulness of one's work becomes the justification for low or no payment — the fact that we need to make a living is completely obscured (I will return to this point shortly). Some sociologists call this practice 'sacrificial labor,' a voluntary economic discount on account of enjoyment in one's labor.[7] This perspective neglects the fact that the practice of sacrificial labor is nevertheless available only to those who can afford it.

I prefer the viewpoint established by the Marxist feminists and their seminal critique of unpaid domestic/reproductive labor, which I expand to the arts via the notion of 'invisiblized labor'.[8] Considering *art work* (the activity, *not* an artwork as in the object) as a form of invisible labor enables us to grasp the mechanisms that produce economically discounted and exploited labor.[9] It also addresses the elephant in the room, namely that 'in a society where access to monetary income continues to be the condition of survival, delivering to capital our work without any compensation is a sure defeat'.[10] Historical analysis of art workers[11] and houseworkers is indeed a formidable witness that unpaid labor has thus far not liberated artists, quite the contrary free labor puts art workers in a position in which they have to accept any job and any pay.[12] Indeed, the power in the field of art production is not wielded by artists but rather by agents who control institutional power and financial resources in educational institutions where they nurture the ideology of creativity and artistic genius, and by art/cultural organizations where this ideology turns into business model based on free labor.

Nonetheless, not all art institutions are on the dark side. In the globally normalized state of the 'free labor business model' according to which museums and galleries largely function, this particular *kunsthaus* and the exhibition *24/7* should be considered a positive example. The artists in this exhibition could wear the overall produced by a multidisciplinary art collective Apparatus 22, with the following text: 'I've got an

Apparatus 22, *ART IS WORK*, Kunsthalle Wien, 2017

artist fee in this exhibition. You understood art is work.'[13] Every artist in this exhibition received an exhibition fee depending on whether they loaned or adapted an existing work or produced a new work. Those who produced new works for exhibition got their production costs covered, moreover, travel expenses were not conflated with honorariums. The Kunsthaus Graz follows the IG Bildende Kunst's recommendations and guidelines for minimum payment for *art work*, which resulted from decades of art workers' struggle for 'fair payment' in the arts.[14] One-third of the budget for this exhibition was spent on artists' fees and art production. There is no doubt that this is a step towards recognizing the fact that art is labor and that, when they provide this kind of services to museums and galleries, artists are workers who need to be remunerated. And yet the majority of art institutions simply do not abide by any such standards. They rely on and reproduce the business model grounded in free labor while we, the blissful audience, enjoy the thrill of good art.

Why is it so hard to acknowledge art as work? Why do we keep obscuring the labor process and not evaluate it appropriately? Every artist will tell you that an exhibition fee does not cover the labor process of making a work of art. Even when, as at this museum, curators and directors are mindful of the economic conditions and the power dynamics in which they operate, they represent a minority. Why do we tolerate this situation? Are we going to receive less enjoyment if art workers are remunerated for their labor? Does the meaning of artists' labor have to be predicated on such economic imbalance? Why do we reproduce these conditions and make it impossible for the less privileged and economically insecure to ever imagine doing art for a living?

After all, making a living is our mutual predicament. 'For capitalism begins not with the offer of work, but with the imperative to earn a living.'[15] The meaning or the imbalance of work and the variegated worlds of labor today cannot take place without pondering the disparities inherent to the enduring capitalist mode of production. Karl Marx, one of the most important thinkers on the exigencies of work but also workers' emancipation under capitalism, posits in *Capital* that work is 'a condition of human existence independent of all forms of society: it is an eternal natural necessity which mediates the metabolism between (wo)man and nature, and therefore human life itself'.[16] Just as spiders weave their webs and bees build honeycomb cells, so humans work too and realize their purpose in the process. Nevertheless, Marx quickly reminds us that our work under capitalism assumes the form of a commodity, labor-power, or labor-capacity that produces not only any number of useful things (use-values) but is also a value-generating substance.[17]

While Marx's argument about labor as the principal value-generating substance in capitalism is central to our understanding of work, he limited it to the sphere of production. Feminist critique of invisible reproductive labor, however, has highlighted a few significant and missing facts. Patriarchal imperial blinders often prevent us from seeing the wageless world upon which the paid or waged labor rests. Put differently, there are several forms of labor — the quintessential one being the work of social reproduction, which refers to all activities that regenerate life, subsistence, and capitalist relations — that don't count as productive and are economically invisible[18] even though they are 'capital's value-generating substance'.[19] Equally significant, the process of value generation for which our labor is central takes place under various forms of exploitation, which occur whether one receives a wage or not.[20] In simple terms, under capitalism, no form of labor is free and beyond the reach of becoming a value-generating substance and consequently also subject to exploitation even if it appears as non-work.

You might say that this does not apply to art production and public art organizations because they are not capitalist firms, but are instead nonprofit organizations. People working in and for art organizations might find their work meaningful, they may even love it, but they exist in the structures that are defined and dominated by a compulsion to earn a living imposed on us by capitalist relations and the capitalist thirst for accumulation. The purpose of art might not be to make a profit, but does that mean that artists and everyone else involved in art production falls outside the exigencies of capitalist relations? How, then, does this special commodity, which 'possesses the peculiar property of being a source of value'[21] operate in the context of art in capitalist society? The subtitle of this exhibition sums it up quite pertinently, as I have already pointed out. But let me offer another example. Whether a urinal is a valuable work of art or not, rests on the fact that a group of artists devised and presented it as such, a museum exhibited it, critics wrote about it, and audiences admired it. Art is indeed a collective practice, as is its aesthetic value generation. Many are involved, but once the symbolic capital is converted into economic parameters, artists are shortchanged. Who (even today) benefits from the value of the most famous urinal? The art workers, or the exorbitantly paid top employees of the museum? Based on the reported facts,[22] it is a straightforward but rich imbalance. Is that the way we in the arts make sense of the imbalance that stems from the meaningfulness of our work?

Lisa Großkopf
→ S. 24 / p. 25

As much as some wishfully profess the end of work at the hands of techno-utopian change, such as automation, robotization, genetic engineering, and the like, the global Covid-19 pandemic has blatantly demonstrated that there will be no end of work under capitalism. Since then, we even have a new distinction between 'essential' and 'non-essential' workers and work. And it should not be lost on us that it is precisely the informalized invisible workers of social reproduction who became visible as the 'global housework'[23] during the pandemic. These are the workers who make capitalism possible not despite their wageless condition, but precisely because of it. And while, in the grand scheme of things, the imbalance related to work rests on the wageless informalized work of the majority of the world population, let's return to our small albeit global world of art and begin to grapple with the free labor business model and the countless wageless hours of art workers who are indeed the largest subsidy art institutions continue to receive.

Notwithstanding the power relations of the global world of art work (or should we say the global art labor market), there is still something to be said about the power that we hold as a collective capable of critical thinking and practice. It is true that individually we are powerless as competitors in the labor market. But if we begin to practice solidarity to de-naturalize the exploitation involved in the unpaid labor from our positions of power — since we all have some — we can collectively change the power relations. I remain convinced that recognizing an artist as a worker and art work as a form of labor, like housework or any other form of work that is value-generating and contributes to capitalist accumulation, is a necessary step if we want to stop giving free gifts to capital — and to see art as a terrain of struggle for a form of life and economy that is based on the value of human needs and work, rather than on value-generating for profit.

Footnotes

1 See Pierre Bourdieu, *Rules of Art: Genesis and Structure of the Literary Field*, Stanford 1992.
2 See Gregory Sholette, *Dark Matter: Art and Politics in the Age of Enterprise Culture*, New York 2011.
3 See Carlo Rovelli, *Reality Is Not What It Seems: The Journey to Quantum Gravity*, New York 2017.
4 See Sholette, 2011.
5 Dieter Lesage, 'A Portrait of the Artist as a Worker', in: Maska 20, no. 94—95: 93, 2005.
6 See Jaka Primorac, Valerija Barada, Edgar Buršić, 'Creative Workers in Permanent Crisis: Labor in the Croatia's Contemporary Arts and Culture', in: Ilya Kiriya, Panos Kompatsiaris, Yiannis Mylonas (eds.), *Industrialization of Creativity and Its Limits*, Cham 2020; also Vahida Ramujkić, Milan Đorđević (eds.), *Rad u umetnosti Rad u umetnosti. Zbornik udruženja likovnih umetnika Srbije* [Work in Art: An Edited Volume of the Association of Fine Artists of Serbia], Belgrade 2021.
7 See Andrew Ross, 'The Mental Labor Problem', in: Social Text 18, no. 2: 1—31, 2000.
8 See Katja Praznik, 'Lexicon: Invisiblized Labor', in: Noah Brehmer, Vaida Stepanovaite (eds.) *Paths to Autonomy*, Vilnius 2022.
9 See Katja Praznik, *Art Work: Invisible Labour and the Legacy of Yugoslav Socialism*, Toronto 2021.
10 See Silvia Federici, *Patriarchy of the Wage: Notes on Marx, Gender and Feminism*, Oakland 2021.
11 See Praznik, 2021.
12 Silvia Federici, 2021, p. 10.
13 NICC Vitrine, Apparatus 22 — 'Art is Work', NICC, February 25, 2022 https://nicc.be/project/apparatus-22-art-is-work/ (14.06.2024).
14 IG Bildende Kunst, 'Guidelines for Fair Remuneration in the Visual Arts', Vienna 2024, https://igbildendekunst.at/en/factsheets/remuneration/ (14.06.2024).
15 Michael Denning, 'Wageless Life', in: New Left Review 66 (Nov—Dec): 79—97, 2010.
16 Karl Marx, *Capital: A Critique of Political Economy, Vol. 1*. New York 1990, p. 133.
17 Ibid., p. 283.
18 See Mariarosa Dalla Costa, Selma James, *The Power of Women and the Subversion of the Community*, Bristol 1973; Silvia Federici, *Wages against Housework*, Bristol 1975; Maria Mies, 1986. *Patriarchy and Accumulation on a World Scale*, London 1986.
19 Alessandra Mezzadri, 'Value theories in motion: Circular labour migration, unfinished land dispossession and reproductive struggles across the urban—rural divide', in: Environment and Planning F 0, no 0 (January 26, 2024), https://doi.org/10.1177/26349825231224027.
20 See Jairus Banaji, *Theory as History: Essays on Modes of Production and Exploitation*, Brill 2010; and Alessandra Mezzadri, ibid. 2024.
21 Marx, 1990, p. 270.
22 Philip Boroff, 'Museum of Modern Art Reveals Glenn Lowry's Whopping $2.1 Million Pay Amid Staff Protest Over Benefit Cuts', in: Artnet, June 4, 2015. https://news.artnet.com/art-world/glenn-lowrys-2-1-million-package-revealed-304820 (14.06.2024).
23 Alessandra Mezzadri, 'The Social Reproduction of Pandemic Surplus Populations and Global Development Narratives on Inequality and Informal Labour', in: Development and Change 53, no. 6: 1230—1253, 2022.

Luiza Margan
→ S. 40 / p. 41

Manuel Correa & Marina Otero Verzier
→ S. 10 / p. 11

Luiza Margan
→ S. 40 / p. 41

Theresa Hattinger, Michael Hieslmair, Michael Zinganel ↑
→ S. 28 / p. 29

Das Museum als Ort aktueller Debatten über „Arbeit“

Zur Schnittmenge von *24/7* im Kunsthaus Graz und *Alles Arbeit* im Museum für Geschichte

Eva Tropper

Es war vielleicht mehr als nur Zufall, dass im Frühjahr 2023 fast Tür an Tür zwei Ausstellungsprojekte geplant wurden, die aktuelle Debatten über ‚Arbeit' zum Ausgangspunkt nehmen. Neben *24/7* im Kunsthaus Graz war das die Ausstellung *Alles Arbeit. Frauen zwischen Erwerbs- und Sorgetätigkeit, Fotoarchiv Blaschka 1950–1966* im Museum für Geschichte.

Als wir als Kurator*innen von unseren gegenseitigen Plänen hörten, fielen uns gleich die Schnittmengen auf: In beiden Projekten ging es, wenn auch aus unterschiedlichen Perspektiven, darum, neu und anders über „Arbeit“ nachzudenken und damit das Museum als Ort für eine höchst virulente gesellschaftliche Debatte zu öffnen. Die Funktion des Museums als ein sozialer Ort, an dem Themen gemeinsam ausverhandelt werden können, war ein wesentlicher Rahmen für beide Ausstellungsprojekte – ebenso wie die Überzeugung, dass Ausstellungen ein hohes Potenzial haben, gesellschaftliche Macht- und Unterdrückungsstrukturen in unserer Arbeitswelt sichtbar zu machen und öffentlich zur Diskussion zu stellen: Wie steht es um gesellschaftliche Verteilungsstrukturen von Arbeit? Wessen Arbeit wird wertgeschätzt? Wem wird Würde, Respekt und faire Behandlung zuteil, wem nicht?

Was wird medial sichtbar und bekommt Aufmerksamkeit, was bleibt unsichtbar?

Während solche Fragen in der Ausstellung *24/7* durch die Linse zeitgenössischer Kunst reflektiert werden, sind in *Alles Arbeit* historische Pressefotografien der Ausgangspunkt einer kritischen Reflexion, die diese Fragen insbesondere im Zusammenhang mit „Geschlecht" aufnimmt und vertieft. Dabei haben uns in *Alles Arbeit* die Fotografien aus den 1950er- und 1960er-Jahren gewissermaßen als Belege gesellschaftlicher Diskurse interessiert, zu denen wir in kritische Distanz kommen wollten. Denn gerade die hartnäckigen Verteilungsungerechtigkeiten im Zusammenhang mit bezahlter und unbezahlter Arbeit basieren auf einer langen Geschichte vergeschlechtlichter Zuschreibungen, an denen Bilder und mediale Repräsentationen einen zentralen Anteil hatten. Bis heute gilt Care- oder Sorgearbeit als überwiegend weibliche Ressource, wie etwa auch die letzte Zeitverwendungsstudie der Statistik Austria eindrucksvoll belegt hat.

In der theoretischen Debatte ist Care- oder Sorgearbeit in letzter Zeit verstärkt in den Blick geraten — als Basis eines Systems von Marktwirtschaft, das diese zwar voraussetzt, aber weitgehend aus dem „Arbeitsparadigma" ausschließt. Der geläufige Arbeitsbegriff — und zwar sowohl der alltägliche, im Satz „Ich gehe arbeiten" gemeinte, als auch der Arbeitsbegriff der ökonomischen Theoriebildung — zielt vor allem auf Formen von Erwerbstätigkeit — trotz der seit der Zweiten Frauenbewegung unter unterschiedlichen Schlagworten geführten Debatte rund um unbezahlte häusliche Arbeit bzw. „Reproduktionsarbeit" und des damit verbundenen politischen Projekts, diese „als Arbeit'" sichtbar zu machen. Umso dringlicher erscheint es, die Diskussion weiterzuführen. Seit den 2000ern, verstärkt aber seit der Corona-Pandemie ist die Debatte wieder aufgenommen worden — angesichts der fortschreitenden Ökonomisierung von Care-Arbeit im Rahmen neoliberaler Strukturen nun unter veränderten Vorzeichen. Diskutiert werden heute Tätigkeiten des Sorgens und Pflegens in unterschiedlichen, unbezahlten ebenso wie prekär bezahlten Kontexten.

Wie aber ist das Thema bisher im Museum angekommen? Eine künstlerische Auseinandersetzung mit unbezahlter häuslicher Arbeit ist seit den 1970er-Jahren wesentlicher Ansatzpunkt feministischer Kunstproduktion — das belegen unter anderem Positionen wie *Semiotics of the Kitchen* von Martha Rosler oder Selma Selmans *AEG Vampyr 1400* in der Ausstellung *24/7*. In kulturhistorischen Ausstellungen sind vor allem in letzter Zeit Zugänge zu verzeichnen, die Sorgearbeit als gesellschaftlich relevantes Thema in den Mittelpunkt rücken. Beispiele sind etwa die Beschäftigung mit Sorgearbeit in *Pflege das Leben. Betreuung* Pflege* Sorgekultur* 2017/18 im Frauenmuseum Hittisau, die Thematisierung von 24-Stunden-Betreuung in *HILFSLINIEN / LINIJE POMOČI* 2020/21 im Pavelhaus oder die Auseinandersetzung mit Fragen der Repräsentation und (Un-)Sichtbarkeit von Sorgearbeit im Projekt *Hoch die Lappen — Feminist Dwellings* vom Sekretariat für Geister, Archivpolitiken und Lücken 2022, in dem das Bildarchiv der *Arbeiter-Zeitung* zum Ausgangspunkt einer Suche nach (raren) Bildern des Sorgens und Pflegens wurde.

Dabei wurden einerseits künstlerische Positionen in die Argumentation kulturhistorischer Ausstellungen aufgenommen. Andererseits gewannen Praxen künstlerischer Recherche an Wichtigkeit, um diese gesellschaftlich so virulenten Themen auf unterschiedlichen Ebenen zu vermitteln.

Auch uns war es ein Anliegen — spätestens seit uns die Schnittmengen der geplanten Ausstellungen *24/7* und *Alles Arbeit* bewusst geworden waren —, einen Zugang künstlerischer Recherche einzubinden, der in beiden Ausstellungen zu sehen sein sollte. Dazu bot es sich an, das Archiv an historischen Pressefotografien der Grazer Agentur „Foto Blaschka", das Grundlage der Ausstellung *Alles Arbeit* ist, für einen solchen Prozess zu öffnen. Gewinnen konnten wir Lia Sudermann und Simon Nagy, die uns von ihrer Beschäftigung mit unsichtbarer Sorgearbeit in ihrem Essay-Film *Invisible hands* bekannt waren. Die daraus
→ S. 62
entstandene Video-Arbeit *Kein Wunder* verknüpft dabei in ihrer gegenwartsbezogenen Fragehaltung die Themen beider Ausstellungen und öffnet sie für eine wichtige Diskussion, die nicht nur, aber auch im Museum zu führen ist.

Literatur

Richard Sandell, Suzanne MacLeod, Ceciel Brouwer, Cesare Cuzzola, *Das Museum als gesellschaftlicher Akteur*, Research Centre for Museums and Galleries, University of Leicester, September 2020.

Katrine Marçal, *Who Cooked Adam Smith's Dinner?: A Story of Women and Economics*, London 2015.

Gabriele Winker, *Care-Revolution. Schritte in eine solidarische Gesellschaft*, Bielefeld 2015.

Eva Tropper, Bettina Habsburg (Hg.), *Alles Arbeit. Frauen zwischen Erwerbs- und Sorgetätigkeit, Fotoarchiv Blaschka 1950—1966*, Ausstellungskatalog, Museum für Geschichte, Graz 2024.

KURS →→
→ S. 36 p. 37

Peter Kogler →→
→ S. 34 p. 35

Loss of motivation
for physical labor

ACEDIA
LISTLESSNESS
SUPINENESS
SLOTH
foot the prejudices of Christian
thought ethics. It must return
proclaim the Rights of Laziness."
…ideas that claim idleness is a sin and diligence is a
were not part of humankind's eternal struggle."
capitalist value system."

The museum as a site for current debates about 'work'

On the overlap between *24/7* at Kunsthaus Graz and *It's All Work* at the History Museum

Eva Tropper

It was perhaps more than just chance that, in the spring of 2023, two exhibition projects exploring current debates about 'work' were being planned almost next door to each other. Alongside *24/7* at the Kunsthaus Graz, this was the exhibition *It's All Work. Women between Paid Employment and Care Work, Fotoarchiv Blaschka 1950-1966* at the History Museum.

When we as curators discovered each other's plans, we immediately recognised the intersections: both projects considered — albeit from different perspectives — how to think about 'work' in a new and different way and so open up the museum as the site for a highly contentious social debate. The function of the museum as a social space, a place where issues can be collectively deliberated, was an essential framework for both exhibition projects — as was the belief that exhibitions hold great potential in terms of identifying the social structures of power and oppression in our working world and putting them in the spotlight for public discussion. What about the social distribution structures of work? Whose work is valued? Who is accorded dignity, respect and fair treatment, and who not? What has visibility in the media and receives attention, and what remains invisible?

While the exhibition *24/7* views questions such as these through the lens of contemporary art, *It's All Work* takes historical press photographs as the starting point for a critical reflection that investigates and delves deeper into these issues with a special focus on 'gender'. In *It's All Work*, we were interested in photographs from the 1950s and 1960s as records of social discourses that we wanted to view from a critical distance. The enduring inequalities in the distribution of paid and unpaid work are built on a long history of gendered attributions in which images and media representations have played a key role. To this day, care work is still seen as a mainly female resource, as strikingly demonstrated in the most recent Time Use Study conducted by Statistics Austria.

In the theoretical debate, care work has recently garnered greater attention — as the basis of a system of market economy that relies on it and yet still largely excludes it from the 'work paradigm'. The common concept of work — both in the everyday meaning of the phrase 'I'm going to work' and the notion of work applied in economic theory — focuses primarily on forms of gainful employment, despite the debate around unpaid domestic labour and 'reproductive work' conducted under various slogans since the second women's movement, and the associated political project of making this visible 'as work'. This makes it all the more urgent that we keep the discussion going. Since the 2000s, and even more so in the wake of the coronavirus pandemic, the debate has resumed — under altered conditions now, given the advancing economisation of care work within the framework of neoliberal structures. Today, there is a discussion around care and nursing activities in various unpaid and precariously paid contexts.

But how has the theme been approached in the museum setting up to now? An artistic examination of unpaid domestic labour has been an important starting point for feminist art production since the 1970s — as is evidenced by positions such as *Semiotics of the Kitchen* by Martha Rosler and Selma Selman's *AEG Vampyr 1400* at the exhibition *24/7*. Cultural history exhibitions in particular have recently presented concepts that highlight care work as a socially relevant topic. Examples include *Pflege das Leben. Betreuung* Pflege* Sorgekultur* 2017/18 at Frauenmuseum Hittisau, the exploration of 24-hour care in *HILFSLINIEN / LINIJE*

PROFIMASTER 2006

POMOČI 2020/21 at the Pavelhaus or tackling questions around the representation and (in)visibility of care work in the project *Hoch die Lappen — Feminist Dwellings* by the Sekretariat für Geister, Archivpolitiken und Lücken in 2022, in which the picture archive of the *Arbeiter-Zeitung* was the starting point in a search for (rare) images of care and nursing.

Artistic positions were incorporated into the argumentation of cultural history exhibitions, while at the same time artistic research practices played a key role in conveying these socially contentious matters on different levels.

It was also vital to us — at least since we became aware of the overlaps between the exhibition plans for *24/7* and *It's All Work* — to integrate an artistic research concept that could be seen in both exhibitions. For this process, it made sense to open up the archive of historical press photographs from the Graz agency 'Foto Blaschka', which forms the basis of the exhibition *It's All Work*. Lia Sudermann and Simon Nagy, whose essay film *Invisible hands* on the theme of invisible care work → p. 63 we knew, agreed to participate. The resulting video work *No wonder* links the themes of the two exhibitions in its contemporary inquiring response, opening them up for an important discussion that should be taken further — both in and beyond the museum context.

Literature

Richard Sandell, Suzanne MacLeod, Ceciel Brouwer, Cesare Cuzzola, *The Case for the Socially Purposeful Museum*, Research Centre for Museums and Galleries, University of Leicester, September 2020.

Katrine Marçal, *Who Cooked Adam Smith's Dinner?: A Story of Women and Economics*, London 2015.

Gabriele Winker, *Care-Revolution. Schritte in eine solidarische Gesellschaft*, Bielefeld 2015.

Eva Tropper, Bettina Habsburg (eds.), *Alles Arbeit. Frauen zwischen Erwerbs- und Sorgetätigkeit, Fotoarchiv Blaschka 1950—1966*, exhibition catalogue, History Museum, Graz 2024.

← Selma Selman
→ S. 58 p. 59

Antje Ehmann & Harun Farocki ↑
→ S. 14 / p. 15, S. 18 / p. 19

Elisa Giardina Papa
→ S. 48 / p. 49

134 bestreikte Stunden*

Simon Nagy

9 to 5
Yeah, they got you where they want you
There's a better life
And you think about it, don't you?
(Dolly Parton, Nine to Five, 1980)

Twenty four seven stay here with me
Twenty four seven throw away the key
Twenty four seven I don't wanna be free
(Tina Turner, Twenty Four Seven, 1999)

In der Ausstellung *No Linear Fucking Time* im Utrechter BAK wird eine berühmte Videoarbeit gezeigt, deren File offenbar beschädigt ist und die deshalb einmal pro Durchlauf kurz hängenbleibt. In dem Hänger steckt eine kleine poetische Kraft, durch die die Idee der Akkumulation von Zeit in der Gegenwart stärker erfahrbar gemacht wird als in der unbeschädigten Version der Arbeit.

→ S. 30

Die Prämisse des „Time Clock Piece" von Tehching Hsieh, das von 1980 bis 1981 produziert wurde, ist schnell erklärt: Der Künstler hat eine Stechuhr in seinem Zimmer installiert, die er sich ein Jahr lang zu jeder vollen Stunde zu betätigen vorgenommen hat. Die Stechuhr ist an eine Kamera gekoppelt, deren Auslöser gedrückt wird und einen Frame ihres Films belichtet, sobald die Stundenkarte gestempelt wird. Neben Hsieh selbst, der sich zu Beginn des Jahres die Haare abrasiert hat, ist auch eine analoge Uhr mit im Bild. Auf den entstehenden Fotos soll die Kombination aus abgebildetem Zifferblatt und sukzessive länger werdenden Haaren den Beweis dafür darstellen, dass wirklich ein Jahr lang und wirklich zu jeder der 24 vollen Stunden des Tages ein Foto aufgenommen wurde.

Die Arbeit wird manchmal auch als Installation ausgestellt. Als solche besteht sie aus 365 Spalten zu je 24 Fotos, die an die Wände gehängt sind. Zeit wird in ihr räumlich, eigentlich tabellarisch abgebildet. Die Fehler, die Hsieh im Laufe des Jahres unterlaufen sind, kommen in dieser Version sehr deutlich hervor: 134 Mal hat er seine konzeptuelle Vorgabe verletzt und den Auslöser nicht zur vollen Stunde betätigt. Die entsprechende Diskrepanz zwischen den 8.760 angestrebten und den 8.626 realen Fotos wird in den 24-Stunden-Spalten sichtbar, von denen einige schlicht um ein oder zwei Fotos kürzer sind als die anderen.
In der Video-Version werden die Fehler nicht so bildlich erfahrbar gemacht, weil die Einzelbilder einander zu schnell ablösen. Das Ausgangsmaterial ist 16-mm-Film, von dem mit jedem Betätigen des Auslösers ein Frame belichtet wird. Die Abspielgeschwindigkeit von 24 Bildern pro Sekunde sorgt dafür, dass jeder Arbeitstag, bestehend aus 24 dokumentierten Stunden, genau eine Sekunde lang zu sehen ist. Die fehlenden 134 Frames knapsen den jeweiligen Tag in Hsiehs Video um je ein Vierundzwanzigstel ab, verkürzen also die Sekunden, die dem Film als Standardzeit innewohnen.

Die Fuck-ups in der Produktion machen das Jahr in der Rezeption um stolze 5 Tage und 13 Stunden kürzer. Würde man sehr genaue Ohren haben und die Arbeit mit Yuri Pattisons Beat abgleichen, der in atomarer Genauigkeit durch das Gebäude vibriert, würde man die leichte Verschiebung irgendwann zu hören beginnen können. Diese Momente des Nicht-Fotografierens lassen sich, wenn man so will, als Momente des Streiks lesen. Obwohl Hsieh in Paratexten versichert, einen extrem lauten Wecker benutzt zu haben, hat er offensichtlich manchmal doch verschlafen. Oder war er sogar das eine oder andere Mal außer Haus, für mehr als eine Stunde am Stück?

Der potenzielle Streik wird in der beschädigten Video-Version, die im BAK gezeigt wird, direkt wieder gebrochen. Und zwar durch den Hänger der beschädigten Videodatei, die Pause im Film. Der Künstler ist beim Wechsel vom einen Frame auf den nächsten auf einmal eingefroren und blickt uns einige Sekunden lang als verschwommene Figur entgegen. Seine unscharfen Konturen lassen ihn ein bisschen gespenstisch wirken.

Während die Vorstellung von bestreikten Stunden die Frage eröffnet, was anstatt des stündlich dokumentierten Arbeitens geschah, macht der knapp fünf Sekunden lange Stillstand eine ganz andere Fantasie auf: nämlich die, dass eine ganz bestimmte Stunde hundertmal so lang ist wie alle anderen. Die Gespensterstunde bekommt ein größeres Fassungsvermögen, nimmt mehr Raum ein. Eine Zeiteinheit dauert schlicht und einfach länger als die anderen. Um ungleich lange Stunden wird es später noch ausführlich gehen, zunächst soll noch der zeitliche Aspekt, der der nicht beschädigten Version der Arbeit innewohnt, beleuchtet werden.

Und zwar lässt sich das „Time Clock Piece" als 24/365-Stück beschreiben. Jonathan Crary hat ein 24/7-Buch geschrieben und macht darin eine gravierende Unterscheidung zwischen diesen beiden Angaben repetitiver Zeit auf. 24/365 impliziert laut ihm eine erweiterte Zeit, in der Dinge passieren können, die eine Weile zum Zubereiten brauchen, also Neues, vielleicht sogar Unvorhersehbares. 24/7 hingegen stellt weniger einen Zeitraum dar als eine statische Wiederholung: das immer selbe und willkürliche Schema einer Woche, das losgelöst ist von der Möglichkeit, Erfahrung oder Entfaltung zu erleben.

Wichtig ist in Crarys Analyse, dass 24/365 durch 24/7 nicht abgelöst wird, sondern Letzteres als zusätzliche Temporalität ins Zeitregime des Kapitalismus eintritt. Die Vorstellung einer unidirektionalen geschichtlichen Entwicklung wird nicht ersetzt, sondern komplementiert von der Loop-Temporalität des 24/7. Die Forderung, lineare Zeit abzuschaffen, lässt sich dadurch in ihrem radikalen Klang hinterfragen. Denn spätestens mit der neoliberalen Konterrevolution führt der Kapitalismus selbst nicht-lineare Zeitlichkeiten in seine Produktionsdynamiken ein. Es geht beides gleichzeitig: die permanente Ausbeutung von Mensch und Natur, die weiterhin auf dem Glauben aufbaut, lineares ökonomisches Wachstum sei mit einem Anstieg von gesellschaftlichem Reichtum gleichzusetzen, und die aktive Segmentierung von Zeit, durch die sie in immer kleinere Einheiten zerteilt wird, in denen sich jeweils nur das Berechenbare wiederholen soll. Mittels unfassbarer Mengen von Forschungsgeldern werden immer neue Wege gesucht, um Decision-Making-Prozesse zu verkürzen, um die für nutzlos erklärte Zeit von Reflexion und Kontemplation auszuschalten oder um Care-Arbeit zu professionalisieren, und das heißt: zu ökonomisieren und Beziehungen zu unterbinden. Die riesige Energie, die in all diese Dinge fließt und die lineare Zeit in immer kleineren Loops um sich selbst kreisen lässt, ist das, was den sogenannten Fortschritt im Kapitalismus darstellt.

* Auszug aus: Simon Nagy, *Zeit abschaffen: Ein hauntologischer Essay gegen die Arbeit, die Familie und die Herrschaft der Zeit*. Münster: Unrast Verlag 2024

Tehching Hsieh
→ S. 30 / p. 31

134 striked hours

Simon Nagy

9 to 5
Yeah, they got you where they want you
There's a better life
And you think about it, don't you?
(Dolly Parton, Nine to Five, 1980)

Twenty four seven stay here with me
Twenty four seven throw away the key
Twenty four seven I don't wanna be free
(Tina Turner, Twenty Four Seven, 1999)

In the exhibition *No Linear Fucking Time* at Utrecht's BAK, a well-known video work is being shown, although the file is clearly damaged and hence it gets stuck every time it plays. There's a poetic force to this sticking, through which the notion of time accumulating in the present becomes starker as an experience than it would be in an undamaged version of the work.

→ p. 31

The premise of 'Time Clock Piece' by Tehching Hsieh, produced from 1980 to 1981, is quick to outline: the artist installed a time clock in his room and committed to operating it at every whole hour for a year. The time clock is attached to a camera, the shutter release of which is pressed, and a frame on its film exposed as soon as the timecard is punched. Alongside Hsieh himself, who shaved off his hair at the beginning of the year, there is also an analogue clock in the image. In the photos that emerge, the combination of the clock-face pictured and the hair successively growing longer are intended to serve as evidence for the fact that, for one year and, indeed, at each of the 24 whole hours of the day, a photo was taken.

The work is also, on occasions, exhibited as an installation. As such, it consists of 365 columns of up to 24 photos each, hanging on the wall. Time is depicted spatially here, or rather in tabular form. The errors that Hsieh made during the course of the year are particularly obvious in this version: there are 134 instances in which he compromised his conceptual guidelines and failed to operate the shutter release on the hour. The corresponding discrepancy between the intended 8,760 photos and the actual 8,626 becomes visible in the 24-hour columns, where some are one or two photos shorter than the others.

In the video version these errors are not made so palpable visually, as the individual photos replace one another too quickly. The base material is 16mm film, one frame of which is exposed each time the shutter is activated. The 24fps playback speed ensures that every working day, consisting of 24 documented hours, is shown for precisely one second. The missing 134 frames rescind one twenty-fourth from each of these days in Hsieh's video, thus making the standard-time seconds inherent to the film that tiny bit shorter.

The fuckups in the production shorten the year in its subsequent reception by a staggering five days and 13 hours. If you had precise enough ears to compare the work with Yuri Pattison's beat, which vibrates through the building with nuclear precision, you'd eventually be able to hear the slight shift. These moments of not-photographing could be read, if you were so inclined, as moments of strike. Although in paratexts Hsieh assures us of having used an extremely loud alarm clock, he obviously slept through it on occasions. Or was he out of the house now and again, for more than an hour at a time?

In the damaged video version shown in BAK, the potential strike is immediately broken by the jarring of the damaged video file, the pause in the film. In the switch from one frame to the next, the artist is suddenly frozen; a blurry figure, he looks back at us for a few seconds. His fuzzy contours make him appear somewhat spectral.

While the idea of striked hours opens up the question of what took place instead of the hourly documented work, the almost five-second standstill evokes a whole other fantasy. Namely, that a very specific hour could be a hundred times longer than all others. The spectral hour gains a greater capacity, takes up more space. One temporal unit just lasts longer than the others.

'Time Clock Piece' can ultimately be described as a 24/365-work. Jonathan Crary wrote a 24/7-book, in which he makes a strong distinction between these two forms of repetitive time. According to him, 24/365 implies a protracted time in which things can happen that need a while to brew — something new, maybe even unanticipated. 24/7, on the other hand, presents less a timeframe than static repetition: the ever same and arbitrary pattern of a week that is detached from the possibility of experience or unfolding.

What is important in Crary's analysis is that 24/365 is not substituted by 24/7. Rather, the latter enters into capitalism's regime of time as an additional temporality. The idea of a unidirectional, historical development is not replaced but complemented by the temporal loop of the 24/7. The demand to abolish linear time can thereby be questioned in its radical tone. Since it was with the neoliberal counterrevolution, at the latest, that capitalism itself introduced non-linear temporalities into its production dynamics. There can be both at the same time: the permanent exploitation of humans and nature, which continues to rest on the belief that linear, economic growth equates with a rise in societal wealth; and the active segmentation of time, through which it is divided into ever smaller units, in each of which only the calculable should reprise itself. Through unimaginable quantities of research funding, new ways to shorten decision-making processes are continually being pursued, in order to eliminate the time for reflexion and contemplation — this having been deemed useless — or to professionalise care work. And this means: to economise and curtail relationships. The monumental energy that flows into all of these things, making linear time revolve in ever shrinking loops, is precisely that which constitutes the so-called progress of capitalism.

Theresa Hattinger, Michael Hieslmair, Michael Zinganel
→ S. 28 / p. 29

Silvio Lorusso
→ S. 38 / p. 39